Microsoft® Planner

Microsoft® Planner

by Jim Boyce

Microsoft® Planner For Dummies®

Contents at a Glance

Table of Contents

Introduction

Welcome to *Microsoft Planner For Dummies*! There are lots of possible reasons why you've chosen this book, but most likely, you're looking for a way to organize and plan something. That *something* may be a set of personal goals, an event, a personal project, or even a business project on which you and others on your team will collaborate. Whatever the result, you may be thinking, "How do I organize all these moving parts into a logical plan?" You've made a good first step in choosing Microsoft Planner!

At a basic level, you can think of Planner as a digital wall or whiteboard where you can arrange sticky notes into a thoughtful plan that helps your project flow from start to successful finish. You can move those pieces of the project around within the plan, add new pieces, remove those that are no longer needed, and do so throughout the project life cycle.

But Planner offers much more than just the capability to move project pieces around on your planning board. Its features enable you to use formatting, color, notes, and other elements to easily visualize and adjust the project whenever needed. Reports and graphs help you track the project elements as you move through the project so you can adjust your approach if needed to stay on track or simply bask in the glory of a well-executed plan.

Your first step was choosing Planner, and you followed that with a second great decision — turning to *Microsoft Planner For Dummies* for help on your journey!

About This Book

This book serves a handful of purposes, but its core purpose is to help you navigate Planner's features, putting those features to work to organize your project and drive it to a successful completion, meeting all your goals along the way. "But wait!" you say. "That's just it. I'm not doing a project. I just want to set some personal goals and work to achieve them." Don't get bogged down by that. You can think of almost anything as a project — a set of tasks that you perform in some sort of sequence to achieve a result. That could be setting and meeting goals for a personal weight loss plan, learning how to play the guitar, or a full-blown project with key deliverables. Whatever the case, Planner can be a great choice for achieving the desired results.

This book isn't just focused on the basics of using Planner, however. It goes beyond the basics to explain how to use graphs and reports to track your project; how to work with Planner in conjunction with other tools, like Microsoft Teams; how to assign tasks to others; and how to use and manage Planner in an enterprise environment.

Much of this book is task-oriented. Most chapters include sets of steps that you perform to learn about and put a feature or concept to work for you. For example, Chapter 6 explains how to create task items within a project board. The chapter includes several step-by-step actions to walk you through the process of creating a task, giving it a name, adding notes to it, and so on. This is an example of a task-oriented chapter.

The book also contains a handful of more conceptual chapters where I explore concepts instead of actions. These chapters help lay the foundation for understanding the types of things you can accomplish with Planner, how Planner compares with other planning tools, the ways in which it integrates with other tools like Teams and Microsoft Power BI, and other supporting concepts that will enable you to really make the most of Planner in any situation.

And of course, all the task-oriented chapters also blend concepts and actions to help reinforce the *why* along with the *how*.

Foolish Assumptions

Every *For Dummies* book makes assumptions about the book's audience, and *Microsoft Planner For Dummies* is no exception. You don't need to be an expert at using planning tools, and you certainly don't need to be a trained or certified project manager to benefit from this book. But I do make a few assumptions.

My first assumption is that you have a good set of basic computer skills and you're comfortable not only using the computer itself, but also using at least a handful of applications. But take heart — you don't need to be a pro at using a computer or lots of applications to use Planner. Basic skills and a desire to organize a plan are all it takes.

When it comes to the chapters that cover integration with Teams (Chapters 5 and 8), my assumption is that you're at least familiar with Teams and use it either for personal use or in an enterprise environment for work. With some of the other tools, like Power BI, I assume less experience or familiarity, so I go into a little more detail in explaining the tool. In a nutshell, I assume a higher level of experience with those tools and applications that most people use often and less experience with the tools that are more specialized.

Chapters 15 and 16 cover topics that are more IT-related and geared toward people who manage IT resources and systems rather than end users. In these chapters, I assume you have not only some IT background but also some experience in managing resources in the enterprise (such as some experience or at least understanding of working in the Microsoft 365 admin center).

Icons Used in This Book

You'll find icons scattered throughout the margins of this book to help you easily identify content that's particularly important. Here's a guide to what the icons mean:

The Tip icon flags practical information that can help you make the most of Planner, get things done more quickly, or use Planner in ways that you may not have considered. These paragraphs often contain advice that you can use right away.

When you see the Remember icon, you know that the information that follows is important. Information in these paragraphs is often conceptual and key to understanding and using Planner effectively.

The Warning icon highlights information that could be detrimental to your success or negatively affect your plans if you ignore it. I don't use this icon much, so don't skip over it when I do. It can help you avoid pitfalls.

When I get into the weeds on a subject, I flag it with the Technical Stuff icon. Feel free to skip these paragraphs if you're just interested in using Planner and you don't need additional technical context (at least for now).

The Premium icon highlights a feature that requires a Premium license for Planner. These licenses include Planner Plan 1, Planner and Project Plan 3, and Planner and Project Plan 5.

Beyond the Book

In addition to the content in this book, *Microsoft Planner For Dummies* includes a Cheat Sheet that you can access online from anywhere using a web browser. In the Cheat Sheet, you'll find helpful tips, shortcuts, and additional content to help you build solid plans and be even more successful in planning and completing projects with the help of Planner. To get to the Cheat Sheet, go to www.dummies.com and type **Microsoft Planner For Dummies Cheat Sheet** in the Search box.

Where to Go from Here

Microsoft Planner For Dummies is packed with information on almost all aspects of Planner. I've laid out the chapters and topics in a logical way that starts with concepts, moves through using Planner, and ends with managing Planner in an enterprise environment. But, like all *For Dummies* books, this book is a reference, so you can pick and choose the bits that you want or need at the moment. If you just want to dive in and start using Planner, for example, you can turn to Chapter 6, and pick up some of the underlying concepts later. Then you may want to flip back to Chapter 5 to learn more about the Planner interface and views.

Or, if you prefer a structured approach, you can start at Chapter 1 and work your way through the book from front to back. There is no right path or right method. Whatever way you approach a topic — whether you skip around or run straight through — your approach will work because it fits *you*. Even after you've finished the book, and in whatever way you choose, you'll likely come back to chapters or sections in a pick-and-choose fashion when you need answers to specific questions.

I'm ready when you are. Let's dive in!

1
Getting Started with Microsoft Planner

Chapter **1**

Introducing Microsoft Planner

The beginning is always a great place to start, isn't it? But maybe you decided to jump ahead to one of the other chapters to dive right into Microsoft Planner and create some tasks. I like to poke around a bit myself before I read the instructions — sometimes until I hit a virtual wall and need help. But you're here now, probably because you want a deeper understanding of what Planner is and in what scenarios it could be useful to you. That's exactly what this chapter is about. I start with an overview of Planner.

Getting Acquainted with Planner

At its core, Planner is a tool that enables you to organize a sequence of tasks within a plan and track those tasks to completion. Planner is incredibly flexible because it provides a framework into which you can fit many different types of plans across a variety of scenarios (see "Investigating Planner Use Cases," later in this chapter). Like many applications, you can use a few of Planner's features to create and manage simple plans, or use all its features — integrated with other applications like Microsoft Teams — to create and manage large or complex projects.

TECHNICAL STUFF

In a nutshell, Teams lets you chat with other users, participate in online meetings, and store and collaboratively work on documents. That's the tip of the iceberg for Teams, because it works with all sorts of other apps, too. If you need more information on what Teams can do, check out Chapter 21 to find resources that explore Teams in detail. Or just search the web for "Microsoft Teams features" — or ask your favorite artificial intelligence (AI) about it.

If you've used the Tasks feature in Microsoft Outlook or the Microsoft To Do app in Microsoft 365, you have some sense of what Planner can do. However, Planner goes well beyond just giving you a means to create and manage a personal task list. With it, you can create plans that organize tasks for yourself and for others into a *project*.

REMEMBER

I use the term *project* to describe any sequence of actions or commitments that are arranged along a timeline and work toward achieving a defined result. This could include establishing a set of personal goals and achieving them, planning a wedding or trip, delivering an educational course, building and executing a business plan, building a house, or developing and delivering an application. All these projects have activities and requirements that span across a timeline and work together toward an end result.

To provide a visual example, Figure 1-1 shows a business plan in Planner, created using the Business Plan template that is included with Planner. The plan is organized into several buckets (more on that topic in the next section), each with some tasks that have checklist items associated with them. Not shown in the figure are additional tasks named Finalize Strategic Plan, Define the Market, and Five-Year Business Plan, each with its own checklists.

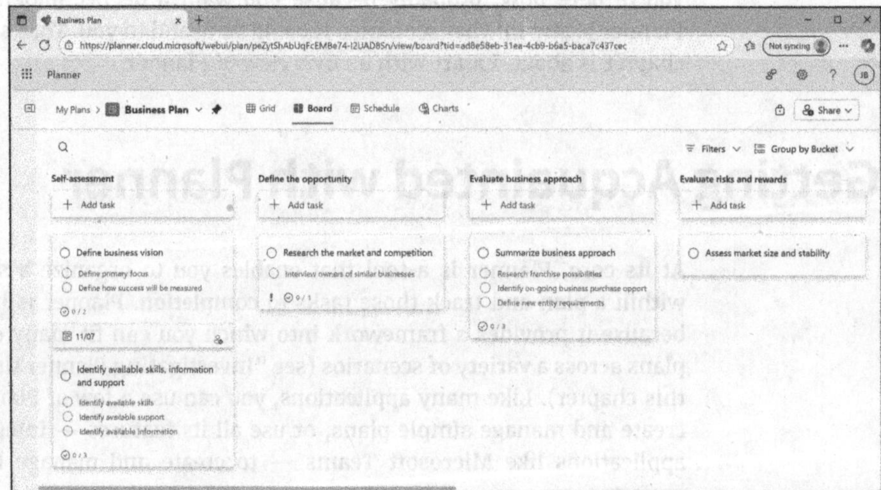

FIGURE 1-1:
Planner includes several templates to help you quickly start a new plan.

Planner certainly isn't the only project-planning tool available today. Several non-Microsoft project-planning tools are available, including Asana, Jira, Trello, and others. I explore some of these alternatives in Chapter 3. In addition, you can think of Planner as an entry point to Microsoft Project, Microsoft's flagship project-planning solution.

Project currently claims about 14 percent of the project-management software market, second after first-place Jira, which holds about 24 percent of the market.

Here's the big picture: Planner offers an easy-to-use tool for planning and tracking tasks and commitments across a timeline as part of an overall plan. There are different versions of Planner — from a free version (Planner Basic Plan), through paid versions, all the way up to Project, each offering increasing levels of features and capabilities. Although other planning tools are available, Planner shines in its integration with Teams and other Microsoft tools. It's also an app published under the Microsoft 365 umbrella of applications, making it readily available to Microsoft 365 users. It's a perfect, inexpensive (or free) tool for project planning across a very broad range of use cases. It's particularly easy for novice planners to use, and it builds nicely toward more feature-rich versions like Project.

At this point, you should have a broad overview of what Planner can do. Other chapters explore these topics in more detail. For example, Chapter 2 digs deeper into the different versions of Planner, how to access those versions, and what points to consider when choosing one. Chapter 3 takes a brief look at some of the other planning tools available today and contrasts them with Planner.

Exploring the Key Features and Benefits of Planner

Planner has a lot to offer, but in this section I focus on the core features for creating and working with tasks, along with the integration possibilities between Planner and other Microsoft tools, including Teams and Power BI.

Capturing and visualizing tasks

The best way to begin to understand what Planner can do is to see it in action. Figure 1-2 shows Planner with a simple plan that contains a handful of tasks for a project called Buy a New Pickup Truck. This is a very basic example because I've only created some tasks; I haven't assigned any start or end dates to sequence them across the project timeline.

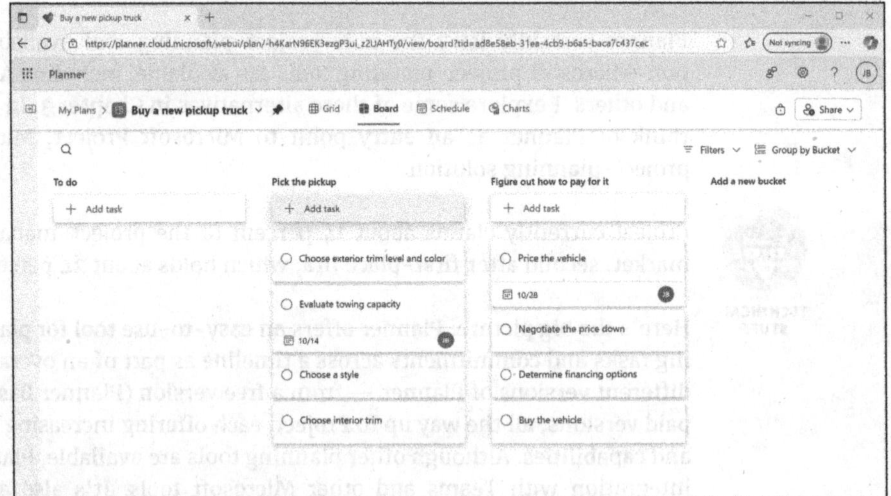

FIGURE 1-2: This is a simple plan for purchasing a new pickup truck.

Organizing tasks

Just slapping sticky notes on a board without any organization may be a good way to get your ideas out, or in this case, start to define key tasks. But any good plan needs to be organized, particularly when a project has *precursors* (tasks that must be completed or requirements that must be met before certain other tasks can begin or be completed) and *dependent items* (tasks that rely on precursors to be done before they can be completed).

Figure 1-2 shows the tasks using the Board view, which works well for visualizing tasks grouped into buckets. I've created a couple of buckets and moved my original tasks from the To Do bucket into buckets that match each task's area, leaving the To Do bucket empty. A bucket doesn't necessarily indicate a sequence for completing tasks. Although I've grouped the tasks in Figure 1-2 into buckets based on two key areas, it doesn't mean that the tasks in the Pick the Pickup bucket need to be completed before the tasks in the Figure Out How to Pay for It bucket.

TIP

Think of buckets as a way to organize tasks into logical groups that generally have something in common but that don't necessarily fall into the same span of the project's timeline.

The Grid view, shown in Figure 1-3, helps you visualize the tasks as a table (referred to as a *grid*). You can edit tasks *inline* in the Grid view, meaning you can edit the tasks' properties right in the grid without opening them in a separate window. The Grid view also lets you filter and sort tasks, for example sorting all tasks by start or end date, showing only the tasks assigned to a specific person, or showing only tasks that have not started yet.

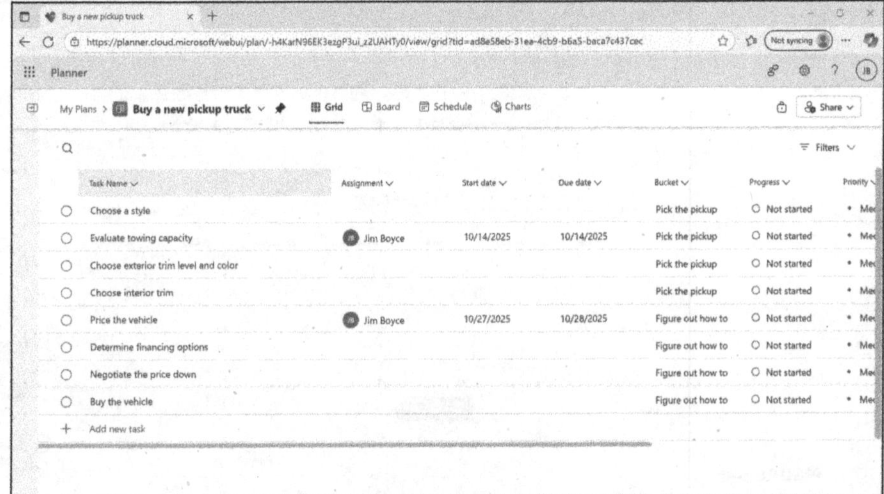

FIGURE 1-3: Use the Grid view to sort and filter all tasks, regardless of what bucket they're in.

TIP

Have a look at Chapter 5 for details on using Planner views effectively. See Chapter 7 for tips on buckets and other ways you can organize and structure tasks.

Scheduling tasks

Every project has a timeline. In almost every case, a task will have a start date and a due date. You can add a start date and a due date to a task when you create it, but in most cases, you'll create tasks, organize them into buckets, and then begin scheduling them after the body of tasks in your plan begins to take shape.

REMEMBER

Some things that you enter as tasks won't have start or end dates. For example, you may include things that you haven't decided are part of the project, or that happen so far in the future that you can't define their dates. Future tasks that do need dates may be dependent on other tasks that haven't been completed yet, so you may leave those dates blank until you get a better idea of when those precursors will be completed. Ongoing tasks like monitoring system performance or responding to customer complaints also won't have clear start or due dates. You could argue that these types of tasks don't belong in a project that has a finite timeline, but adding them helps you remember and plan for those activities. You could group them together into a bucket called Things That Never End!

This is a good place to introduce the Schedule view, shown in Figure 1-4. This view shows your tasks by due date on a calendar to help you visualize not only when the task is supposed to end, but also how it relates to other tasks on the timeline.

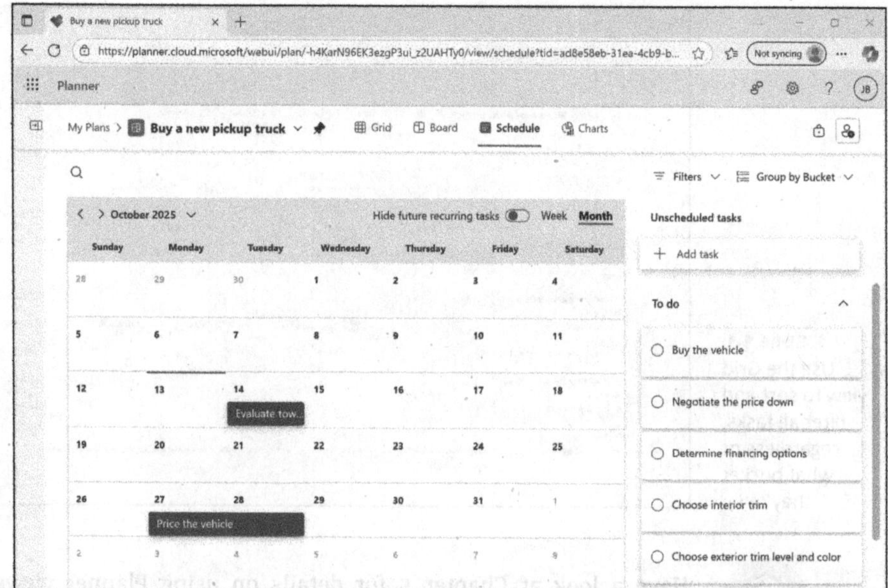

FIGURE 1-4:
Use the Schedule
view to visualize
tasks on
a calendar.

Assigning tasks

Planner is a great tool for creating, managing, and completing personal projects. But it's also a great tool for teams of people to manage and deliver projects. The key feature that supports teams in Planner is task assignment. Figure 1-5 shows a task that I've assigned to myself, indicated by my name near the upper left of the task form. I need to check the axle ratio and other specifications of the pickup I want to buy to determine how much weight I can haul with it.

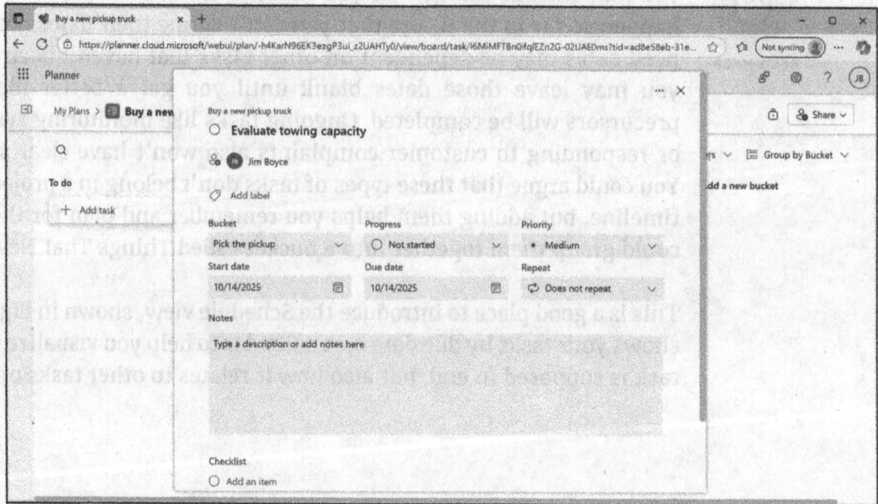

FIGURE 1-5:
I've assigned a
task to myself.

Assigning a task is easy! For example, in the Grid view, just click the Assignment column and pick a person. You can also assign tasks in the Board view. The assignee receives a notification after you assign a task to them.

TIP There are a few requirements for assigning tasks to someone, one of which is that they have to be a member of the plan. For example, people in your organization must have a valid Microsoft 365 account in your organization or tenant, and external users must be added as guests to the group. Check out Chapter 9 to find out more about those requirements.

Tracking tasks

Planner isn't just for building a plan and creating tasks. It's also the tool you use to track how your plan is progressing. You probably want to know how many tasks are complete, who completed them, how many are left, and so on. Figure 1-6 shows the Charts view, which gives you that type of information.

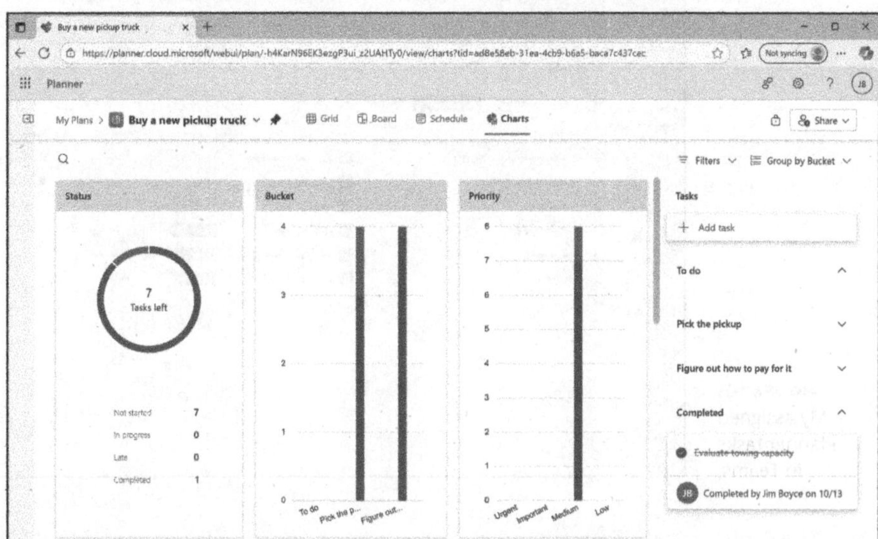

FIGURE 1-6: The Charts view provides information about your plan's progress.

I admit that the information in the Charts view is a bit rudimentary, but even so, it offers a concise view of your project's status. There are some other ways to visualize progress that requires exporting the data out of the plan. Check out Chapter 15 to find out more.

TIP

Microsoft Planner For Dummies covers the Planner Basic Plan and Planner Plan 1. The Basic Plan is limited to the reporting features provided by the Charts view (unless you export the data for visualization in another application). Plan 1 adds a Timeline view, which is a Gantt chart that shows task duration along a timeline. Plan 1 and other versions, explored in more detail in Chapter 2, offer more features for charting, visualizing, and reporting.

Integrating with other applications

Planner is a great tool for organizing your personal projects, but it's also great for projects that you plan and deliver with a team of other people. I spend most of my day in Outlook and Teams, collaborating with coworkers as well as customers. Planner offers integration features with other applications, including Teams and Outlook. Figure 1-7 shows an example of the Planner app in Teams, where I'm viewing the tasks assigned to me.

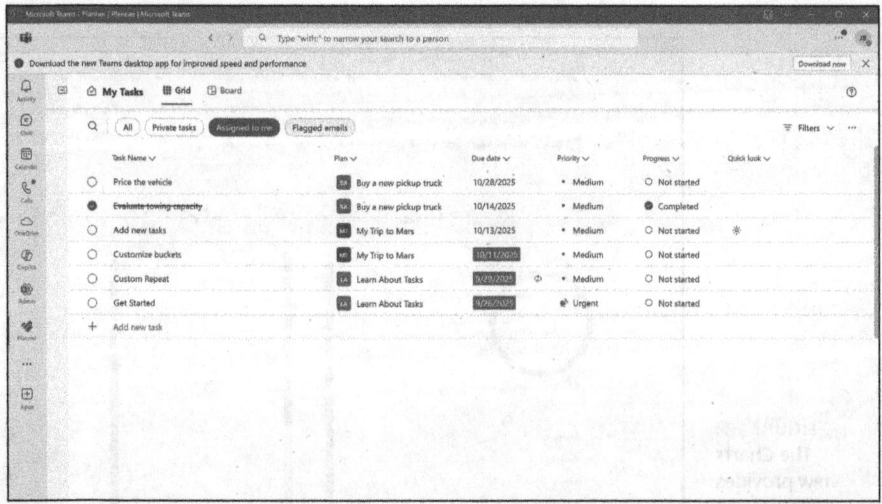

FIGURE 1-7:
My assigned
Planner tasks
in Teams.

Using Planner in Teams is just like using it in a web browser, which means that if you work in Teams a lot, you don't have to open a separate window to manage your tasks. Planner also integrates in other ways with Teams, including the capability to add Planner as a tab in a team, receive task notifications in Teams, and integrate Planner and Microsoft SharePoint for plans that are linked to Teams sites.

In a nutshell, SharePoint is a solution for creating and publishing websites that enable organizations to share documents, publish web content, and collaborate in other ways. SharePoint also provides tools for workflow automation (like document-approval processes), content management and governance, storing information in SharePoint lists, and much more. For more information on Share-Point, Chapter 19 is a good place to start. Or search for "Microsoft SharePoint features" using your favorite search engine or AI tool.

Integrating Planner with Teams and Outlook isn't particularly difficult, but you do need to set it up. When you're ready to do that, turn to Chapter 8.

Investigating Planner Use Cases

Now that you have a sense of what Planner can do, you're ready to explore some of the use cases where Planner can be a great planning and task management tool. You may already have ideas for what you want to use Planner for, but the following sections can help you expand your horizons and maybe integrate Planner into your personal life, as well as work.

Not familiar with the term *use case*? It's simple. A *use case* describes how a user or system interacts with a product, process, or service to reach a goal. Think of a use case as describing how you may use Planner in different situations to achieve different types of goals. I like the hammer analogy: You can use a hammer to pound in nails or pry them out, pry boards apart, open walnuts and coconuts, pretend it's a microphone, or vent your frustrations (only on inanimate objects like coconuts, please). All those are use cases for a hammer. Nailed it!

Personal achievement

Most of us, whether we externalize them or not, have personal goals that we want to achieve. I could stand to lose about 40 pounds of fat and gain 10 pounds of muscle. But I probably won't be successful without a plan that includes changing what I eat, exercising, getting lots of sleep, and hydrating. That plan should include a regular exercise schedule, instead of just a task called Exercise!

Don't take this as fitness advice, but I should combine cardio, weight training, and endurance exercises into my plan, scheduling them each week. Maybe Monday should be lower-body training, Tuesday high-intensity cardio, Wednesday upper-body training, Thursday walking, Friday abs day, and Saturday more high-intensity cardio. Sunday would, of course, be lying-on-the-couch day, because now I can barely move. (Maybe you can understand why I need to lose some weight.)

I'm an overweight, out-of-shape IT geek and author, so don't take fitness advice from me! Consult your doctor before you start any kind of weight-loss or exercise plan. While you're there, get their advice on a good fitness plan that you can build into Planner. I bet they even have a pamphlet they can give you to get started!

Here are some examples of personal achievement use cases for Planner:

>> **Health and fitness:** Weight loss is just one scenario where you may want to create a personal plan. Maybe you want to learn or get better at yoga, gain some muscle, run a marathon, or get washboard abs. All these goals deserve a plan.

>> **Career advancement:** This scenario may take the form of achieving a work-related certification, changing roles, earning a promotion, becoming a leader or people manager, or expanding your network. Chart a path to that goal with a good plan.

>> **Financial milestones:** Remember that new pickup truck I wanted to buy earlier in this chapter? Great example, if I do say so myself! Saving to buy a house, paying off a mortgage, paying down debt, and building or expanding your investment or retirement plans are all good examples of planning for financial milestones that need a plan that fits your goals.

>> **Education:** This could encompass a work-related certification, working toward a degree, learning a new skill or musical instrument, or upskilling in your favorite hobbies.

>> **Personal skills development:** Do you want to improve your public speaking skills without relying on written speeches? That's a great example of personal skills development. Improving your time management abilities, gaining leadership skills, and improving your writing skills are all good examples of personal skills development.

Course planning

I spent seven years as an instructor at a technical college teaching mechanical drafting, programming, and computer-aided design. I did a lot of course planning during that time and have done some additional course-planning delivery in other roles, including the one I'm in now. You wouldn't walk into a classroom on day one of a quarter or semester without a plan! That plan needs to detail the syllabus, the learning objectives to cover, what testing measures you'll use throughout the course, and what the final assessment will be to make sure your students have achieved an appropriate level of success in the course.

Planning a course can be a bit different from planning an event or "traditional" project. For example, each task can describe a single lesson plan, using the due date as the date on which you finish delivering the lesson. Use buckets to describe the course objectives and place lesson-plan tasks in their respective buckets. Use notes to describe the lesson plan and any resources or references required for delivering it. You can also attach Uniform Resource Locator (URL) links (otherwise known as website addresses) or documents to a task to incorporate external content into the lesson.

You can insert a link in a task regardless of which version of Planner you're using. Plans that are integrated with Teams add the capability to attach documents from SharePoint libraries associated with the team. If your plan wasn't created within a Microsoft 365 group or in Teams, but you still want to link to a document stored in SharePoint or some other location, open the document in a browser, copy the URL, and then insert that URL in your task.

"What about assignments?" you ask. You can include student tasks in your course plan and assign them to your students. Unfortunately, Planner supports assignment of a single task to a maximum of 20 people. If your class head count is 20 or fewer, you're in luck. However, keep in mind that if only one of your students marks the task as complete, it's complete and disappears from the My Tasks view for all students.

A task in Planner is a single shared item. Even if standard task assignment in Planner isn't geared toward individual completion by your students, it can still be useful. For example, you could divide the students into smaller groups, each responsible for a particular task within the plan. When the group finishes the task, they can mark it as complete *for the group.*

So, assignment is a tough nut to crack for this use case. Here are some suggestions on how to overcome this limitation:

>> **Manually copy the task.** You can copy the task *n* number of times, one for each student, and then assign them individually. That's a lot of work with a large class!

>> **Automate task creation with Microsoft Power Automate.** You can build a flow using Power Automate that loops through a list of student names and creates and assigns a task to each one. That's a great solution, but it requires a little bit of setup and an understanding of how to use Power Automate (see Chapter 13).

>> **Consider an alternative solution.** Planner may not be the best choice if you have a large class and want to assign tasks to each student. A learning management system (LMS) or other solution may work better for that aspect of the course. Or you could still build your course plan in Planner, but use a complementary solution for assigning tasks.

Event planning

This use case is straightforward: You want to plan a wedding, birthday party, or other big event. I use a wedding in this example, which would likely include the following bodies of work:

>> Budget

>> Ceremony location

>> Reception venue

>> Weather backup plan

>> Guest list

>> Caterer

>> Photographer and videographer

>> Attire

>> Décor

>> Vendors

>> Invitations and stationery

>> Music

>> Legal stuff

There is no right or wrong way to structure a wedding plan in Planner, but here are some suggestions (keeping in mind that I know Planner, but I am in no way, shape, or form a wedding planner):

>> **Use buckets to organize the various bodies of work.** Each of the items I list earlier would have its own bucket, with additional buckets added for the things I (or you) haven't thought of yet. Within the Venue bucket, for example, lay out the tasks necessary for choosing the venue, pricing it, reserving it, decorating and cleaning it, and so on.

> » **Rely on checklists.** Use checklists to define the individual actions that need to be completed within the overall task.
>
> » **Include contingencies in each task or bucket.** Something will inevitably go wrong with the plan, so build in contingencies at the task level, bucket level, or both.

Project planning

Some people may think of this use case as the primary one for Planner, but I see it as one of many — but also one that covers many scenarios. I use *project planning* in this context to mean a project that delivers a tangible product or service in a business environment. This could include building or renovating a house, developing and delivering a software application, performing a data-center migration, launching a new product, and much more.

REMEMBER

I'll admit that project planning can encompass *anything* that needs a plan. So, all the use cases I describe in this chapter are examples of project planning. But I include this use case to contrast projects that deliver tangible *things* from those that deliver an *end result*.

Although this use case might be the most common, it's also the one with the most variability in how you build the plan. That's because it covers so many possibilities. The plan for launching a new software application is very different from the plan for building a house. Even so, all these projects can use buckets to organize tasks, have tasks with start and end dates, need task owners, and potentially include open-ended tasks that you plan for after the project's completion.

Student projects

Planning an educational course is different from planning a student project, but a student project will often follow or be a key part of a course. For example, assume you're teaching a course on framing for home construction. You'll likely deliver the course as classroom instruction interspersed with hands-on learning. You start with a module on floor systems; then you go out and build the first floor. Next comes a module on wall systems, followed by raising of the first-floor walls. You wouldn't teach the whole course in one go and then go build the house, because by the time you've explained the relevant parts of the residential building code and how to frame the whole house, your students have forgotten the right way to tie the rim joists to the foundation.

In this example, you'd create hands-on sessions as individual plans that you'd link to the overall course plan. The floor system would have one, the wall system another, roof framing a third, and so on. Although you could set up the entire construction phase as a single plan, breaking it into smaller components not only makes it easier to focus on specific modules, but can make it easier to address changes. Breaking it into several smaller plans also makes plan development a less daunting task.

That's the way I would do it in an educational setting, but that's often not the case in the construction trade where you're planning for a project that will be completed by skilled tradespeople. In many cases, a residential construction project is backed by a single, comprehensive master plan that includes all the major systems. More complex projects, such as multiunit housing and larger commercial projects, have a master plan for the entire project with sub-plans for each system (structural, electrical, plumbing, and so on).

Now consider other types of student projects, such as science projects. These are certainly less complex than a building project, but they still need a plan. In this scenario, create a plan that covers the specific science module and include a bucket in that plan for the student's creation of their project.

Here's the key: Have each student create their own plan for their project. This, more than anything, will help reinforce the topic and also help them organize their thoughts and deliverables.

Business planning

I confess that I've jumped into businesses without a plan. For example, I once owned a small internet service provider (ISP) business. I would joke that it didn't go as planned, but I *didn't* plan, which explains why I no longer own an ISP! Whether you're planning to start a home-based business or something that will need a manufacturing facility and employ lots of people, you should start with a solid business plan.

Fortunately, Planner includes a Business Plan template that takes you through a thoughtful analysis and assessment that includes the following:

>> Performing a self-assessment

>> Defining the opportunity

>> Evaluating a business approach

>> Evaluating risks and rewards

>> Finalizing a strategic plan

>> Defining the market

>> Building a five-year business plan

Chapter 11 explains Planner templates in general and explores the various templates that are included with Planner.

Your plan may not include all those buckets, or it may contain more, depending on the size, scope, and purpose of your business. In any case, the business plan serves as the mechanism for you to explore what the business will be, how you'll build and deliver it, what skills you'll need, and so much more. Last, don't skimp on your first five-year plan. If you do, you'll find yourself reacting to your business rather than guiding it as both you and the market change.

Why do I call it your *first* five-year plan? Successful businesses are constantly planning for the future. Even if you're planning a small, hobby-based business, you'll want that business to not only grow and flourish but also adapt to all sorts of potential changes. Planning your business's future over the long term is the best way to ensure long-term success.

Thinking about How You Want to Use Planner

If you've read this chapter from the beginning, you probably have a good sense of how you can put Planner to work in any number of ways. You've probably thought of variations on the use cases I describe and come up with a few of your own. That's the beauty of a good organizational tool like Planner. It doesn't force you into a rigid framework. Instead, it offers the flexibility you need to create any type of plan, map tasks to the plan, and work toward an end goal.

This chapter explores what Planner is and looks at some common scenarios in which you may use it. A good next step at this point is to pause and think about how you want to use Planner, at least for your first project. Start by defining your end goal. Sketch out some buckets that make sense for your project and come up with some tasks for each. Think about how all that drives your project forward to completion. You'll then have a framework in which to better understand the topics in the rest of this book.

Chapter 2

Choosing a Plan and a Path

Unlike many other applications that have a single version with a single set of features and capabilities, Microsoft Planner is available in a handful of versions. These versions build on one another, adding features while also adding cost. If you understand the capabilities of these versions, you'll be able to choose the version that's right for you — or, at the very least, understand the capabilities of the one that's available to you in your organization.

This chapter explores each version of Planner to help you compare features. It also highlights two additional ways to interact with Planner through Microsoft Teams and the Planner web app (both covered in more detail in Chapter 5).

Understanding Plans and Licensing

Microsoft Planner For Dummies focuses on two versions of Planner: Planner (also called Planner Basic) and Planner Plan 1. Even though I don't cover the other versions in detail in this book, it's good to understand the additional features those versions offer so you can determine if a different version's capabilities make it worth upgrading.

Whenever appropriate, I note the feature differences and minimum version needed to achieve the described activity or outcome.

The version names described in this chapter are the official names at the time this is written. The names may change over time. In fact, the names changed as I was writing this book. (To check for updates, see www.microsoft.com/en-us/microsoft-365/planner/microsoft-planner-plans-and-pricing.) Also keep in mind that Planner is in active development, so features may change from those I describe in this book. Licensing requirements could also change for specific features.

The following sections explore the various versions of Planner and how each is licensed, as well as the features in each version.

Planner

Planner, sometimes called Planner Basic, is the entry-level version of Planner. Although I've dubbed it entry-level, that doesn't mean it skimps on features. Planner offers all the features and capabilities you need to create and organize plans, work with multiple views, use reporting, assign tasks, and collaborate with people in other ways on plans and tasks.

Planner is included with most Microsoft 365 license plans, including the following:

>> Microsoft 365 Business Standard

>> Microsoft 365 Business Premium

>> Microsoft 365 Enterprise E1, E3, and E5

>> Microsoft 365 Education A3 and A5

>> Microsoft 365 Government Community Cloud (GCC)

See www.microsoft.com/en-us/microsoft-365/planner/microsoft-planner-plans-and-pricing to learn about Microsoft 365 licensing plans.

If you have one of the Planner-supporting licenses, Planner is free in the sense that you don't have to pay extra for it. A few Microsoft 365 license types do *not* include access to Planner. These license types include the following:

>> Microsoft 365 Business Basic

>> Microsoft 365 F1, F3, and other frontline worker licenses

>> Microsoft 365 Apps for both Business and Enterprise

>> Microsoft Office 365 Education A1

>> Microsoft Teams Phone (if purchased as a stand-alone add-on)

>> Other specialized add-on licenses such as Intune-only or Defender-only licenses

Planner Plan 1

Planner Plan 1 (formerly known as Project Plan 1) includes the following features, in addition to the features in Planner:

>> **Project goals:** Define up to ten top-level project goals and link tasks to those goals. This feature can help the team focus on strategic tasks and objectives to ensure they align to key priorities. (See Chapter 12 for details on working with goals.)

>> **Backlogs and sprints:** A *backlog* is an ordered list of tasks tied to future work. A *sprint* is a time-boxed iteration during which a team works on a set of tasks from the backlog. Planner Plan 1 offers some rudimentary features for managing backlogs and sprints. (See Chapter 12 for details on building sprints.)

>> **Premium plan templates:** A Premium license such as Planner Plan 1 gives you access to several Premium templates that generate prebuilt plans with buckets, tasks, and other items already configured. Some of the Premium templates include Project Management, Commercial Construction, and Sprint Planning, along with several others. (See Chapter 11 for details on Basic and Premium templates in Planner.)

>> **Expanded reports and dashboards:** Planner Plan 1 adds more capabilities for standard reports and dashboards and supports more granular reporting through creation and filtering of custom fields, dependencies, and milestones. (See Chapter 15 for details on creating and fine-tuning reports.)

>> **Timeline view:** This view shows tasks organized on a Gantt chart, enabling you to visualize task sequence across a timeline. (See Chapter 5 for details on using the Timeline view.)

>> **Task dependencies:** A *task dependency* makes a task dependent on another task. You can create Finish-to-Start dependencies between tasks in Planner Plan 1, meaning, for example, that Task A must complete before Task B can begin. Creating dependencies for Start-to-Start, Start-to-Finish, or Finish-to-Finish require Planner and Project Plan 3 or Planner and Project Plan 5. (See Chapter 12 for more on dependencies.)

>> **Microsoft 365 Copilot in Planner:** Use Copilot artificial intelligence (AI) to generate plans, analyze progress and risks, and manage projects in other ways using natural language queries.

>> **Additional customization features:** You can create up to ten custom fields in a plan with Planner Plan 1 using any of several field types, including Text, Date, Number, Yes/No, and Choice. Custom fields can provide clarity and additional detail within a plan; they can also be used to organize and report on tasks in ways not available through the predefined reports. (See Chapter 12 for more on custom fields.)

You can add the Premium features of Planner Plan 1 by purchasing a Planner Plan 1 license, which gives you an annual subscription that renews automatically (unless you or your admin turns off auto-renew).

Planner and Project Plan 3

As with Planner Plan 1, you can purchase an annual subscription for Planner and Project Plan 3. This version of Planner includes all the features in Planner Plan 1 and adds the following:

>> **Task history:** Track task history with auditing details to determine who changed something, what they changed, and so on. Change history is useful not only for auditing but also to help identify and understand delays and other project issues.

>> **Roadmaps:** This feature enables you to view project timelines, key dates, and deliverables across multiple plans. Roadmaps are integrated into the Portfolio feature (described later in this list), enabling you to coordinate and drive a portfolio of projects.

>> **Baselines and critical path:** A *baseline* captures a snapshot of a plan at a specific point in time and is used as a reference point to measure actual plan progress. The *critical path* in a plan is the longest sequence of dependent tasks that determine the earliest possible completion date for the plan. If any task on the critical path is delayed, then the project completion date is also delayed.

>> **Resources request capability:** Planner and Project Plan 3 expands the capabilities for a resource manager to view and manage the assignment of people and other resources to a project.

>> **Program management and Portfolios:** You can view and manage multiple projects together in a portfolio of projects, providing good insight and visualization across multiple projects.

>> **Project financials, budgeting, and costing:** Planner and Project Plan 3 includes features for developing a project budget, tracking estimated versus actual costs, and monitoring and reporting on project financial health.

>> **Advanced dependencies with lead and lag times:** Create dependent tasks with lead (overlap) and lag (delay) between tasks.

>> **Project Online and desktop client:** Access Project Online and use the desktop client to manage projects.

Planner and Project Plan 5

Planner and Project Plan 5 includes all the features in Planner and Project Plan 3 and adds the following:

>> **Advanced portfolio management:** These features include broader reporting across all projects across the organization and expanded analytical capabilities to evaluate projects across a portfolio. You can model different scenarios and determine which projects offer better return on investment and growth based on constraints such as costs, resources, and other factors.

>> **Enterprise resource management (ERM) and allocation:** Gain several features that enable ERM, including viewing resource allocation across portfolios and the entire organization, and use capacity and availability views to manage resource loading across all projects.

Identifying Premium features

As you become familiar with Planner, you'll find that Premium features are usually identified by a diamond icon, as shown in Figure 2-1. Wherever appropriate, an icon in the left margin identifies Premium features throughout this book.

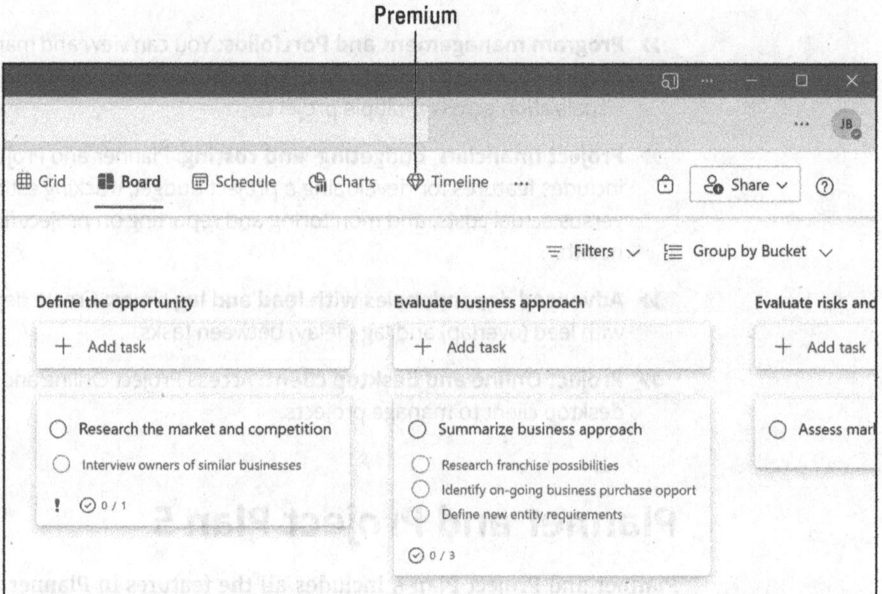

FIGURE 2-1:
Premium
features are
usually marked
by a diamond
icon in Planner.

Accessing Planner

Chapter 5 explores the Planner web interface, integration between Planner and Teams, and the Planner mobile app. You may want to explore some of the Premium features described in this chapter without referencing Chapter 5, so I touch on a couple of relevant topics here. Refer to Chapter 5 for more details.

The Planner web app

You can access Planner at `https://planner.cloud.microsoft`. The web app gives you access to most of Planner's features (see Figure 2-2).

TIP

Although Microsoft will likely integrate all features into the web app, as I write this, some Premium features are available only in Planner through Teams.

TECHNICAL STUFF

"Why are features split up like that?" you ask. When an application like Planner moves to a software as a service (SaaS) model rather than a discrete, published desktop application, the developers have more flexibility in how they design and deliver the solution. They aren't building and distributing a complete application in a box or through a file download. Everything is hosted online, giving them the capability to add features incrementally because the solution can be delivered modularly. In addition, some features are made available in preview, where a feature may be published, in this case, through Teams but not through the web app.

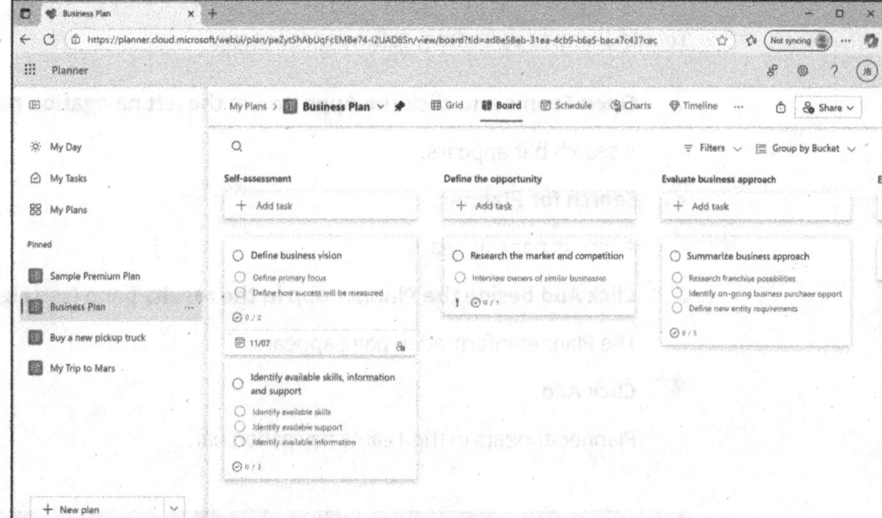

FIGURE 2-2:
You can use the Planner web app to access most features.

Planner in Teams

In addition to using the web app, you can also access Planner in Teams, assuming you've been assigned a Teams license. Figure 2-3 shows Planner in Teams, and as the figure illustrates, there really isn't much difference — at least on the surface — between Planner in Teams and the Planner web app. However, as I explain in the preceding section, some features, at least for the time being, may be available only through Teams.

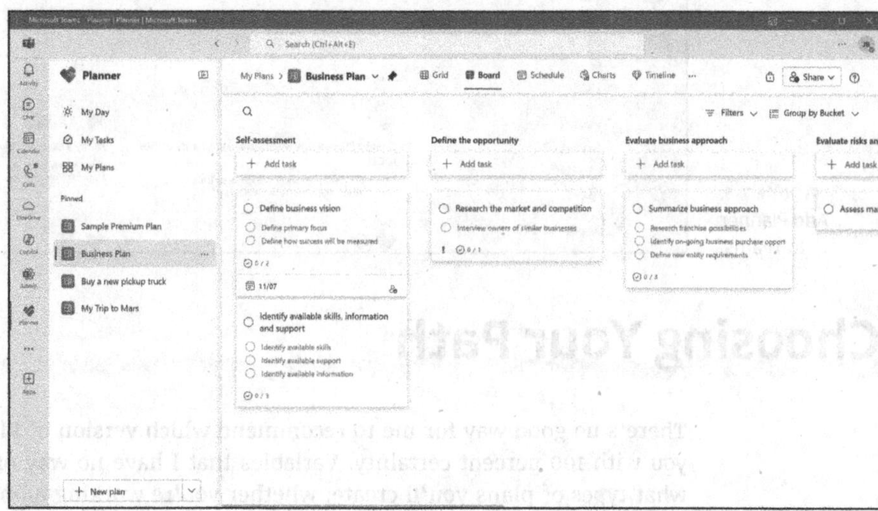

FIGURE 2-3:
You can also access Planner through Teams.

To add Planner to Teams, follow these steps:

1. **Open Teams and click the Apps icon in the left navigation pane.**

   A search bar appears.

2. **Search for Planner.**

   A results pane appears.

3. **Click Add beside the Planner app in the results pane (see Figure 2-4).**

   The Planner information page appears.

4. **Click Add.**

   Planner appears in the Teams navigation bar.

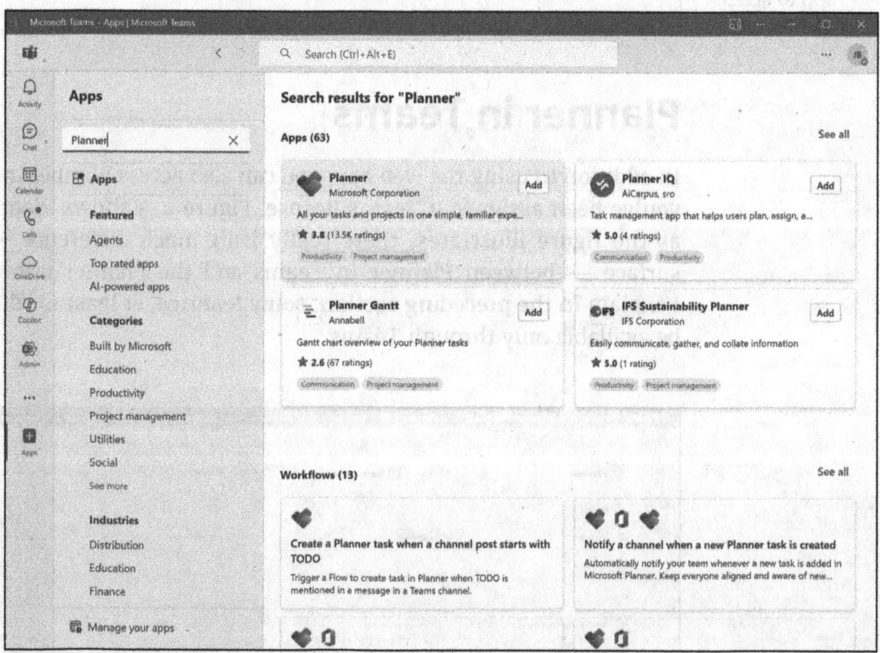

FIGURE 2-4:
Add Planner
to Teams.

Choosing Your Path

There's no good way for me to recommend which version of Planner is right for you with 100 percent certainty. Variables that I have no way of knowing — like what types of plans you'll create, whether you're working alone or with a team, the size of your organization if you're using Planner at work, your comfort level with project management, and a host of other variables — will determine which version best suits your needs. However, I can offer some general guidance to help you choose:

>> For **personal goal setting and planning** that doesn't require features like dependencies or Timeline view, Planner Basic is a good starting point. You can always upgrade to a Premium plan if your needs outgrow the Basic version.

>> **Small teams,** whether in a business or private group, can also benefit from Planner Basic. You can assign and track tasks and collaborate in other ways without the need for a Premium license.

>> If you use Planner in a **business setting,** whether by yourself in a small business or with team members in a larger organization, the Premium features in Planner Plan 1 are compelling. The capabilities to add simple dependencies, use sprints, and leverage Copilot are worth the investment in a Planner Plan 1 license. If the work you do with Planner generates revenue, consider a Premium plan.

>> If you need **portfolio management and enterprise resource management,** choosing Planner and Project Plan 3 or Planner and Project Plan 5 is likely a good investment.

This book covers Planner Basic and Planner Plan 1, but Even if you find somewhere along your Planner journey that you need the features in the other Premium versions, this book still provides a strong foundation in a very broad range of Planner's capabilities. Not only can you create and manage many types of projects both large and small with the features covered in this book, but you can build on this foundation as you move up through more complex projects that do require the features in the other Premium versions.

Chapter 3

Comparing Microsoft Planner to Other Tools

You're reading this book, so it's a sound bet that you've already decided to use Microsoft Planner. Your reasons could vary. Maybe you're using it because it's included with your Microsoft 365 license. Maybe it's the solution that your organization has chosen. Or maybe you've chosen it as a means to become familiar with project management concepts. Regardless of your reasons for choosing Planner, there are alternative solutions available. This chapter covers some of the most popular solutions to help you understand the similarities and differences between those solutions and Planner. Understanding the features in these other solutions can help you increase your overall project management knowledge and better understand the role that Planner plays in the market.

Defining Your Needs

Did you build a plan to learn about project management in general and Planner in particular? I'll hazard to guess no. But if you did, congratulations! You've passed the first hurdle to getting organized. Did you include defining your needs in the plan? That's the true starting point.

I can't tell you what planning and project management features you need because I have no way of knowing how you intend to use Planner, what types of projects you're running, and so on. But I can offer some guidance based on the features available in the different versions of Planner, as well as the features in some of the alternative solutions.

If all you need is the capability to create and work with a to-do list, then Planner Basic has you covered. In fact, the to-do features in Microsoft Outlook or the To Do app in Microsoft 365 may be just as suited to your needs. You won't get the planning and organization features you find in Planner, but you can at least create and check off items in a to-do list. Personally, I've used both of those solutions, and neither one lasted very long in my tool belt before I dropped them in favor of Planner.

The basics

Whether you need to create simple plans for your personal use or complex plans for enterprise projects, the solution you choose needs to offer a good set of basic capabilities:

>> **Creating and assigning tasks:** This is the whole point of any project management tool!

>> **Due dates and reminders:** Any solution needs to give you the capability to assign start and due dates so you can distribute tasks across the plan's timeline and show the timeline relationship between tasks. Notifications are important to help you stay on track, particularly if you're also focused on other responsibilities.

>> **Templates:** Templates can be a major time-saver in setting up plans, particularly for complex ones. They also help by providing built-in guidance on what your plan should include.

>> **Charts and visualizations:** You may not need charts and visualizations in a very simple plan, but they become invaluable as plan complexity grows. You need a good set of tools to visualize progress, challenges, and risks throughout the lifespan of the project. The more complex the projects, the more visualization options, reports, and dashboards you need to fully track success.

You'll find a Premium icon sprinkled through this chapter where a described feature is only available with a Planner Premium license. Where applicable, I note how you can simulate a Premium feature, but the Premium icon represents a built-in Premium feature.

Scheduling and planning

I wrote in the previous section that creating and assigning tasks is the whole point of a planning tool. That's true, but I need to take it one step further. You also need a good set of tools to schedule those tasks. Without that capability, you're just throwing tasks into a bucket with no interrelationships. That's a to-do bucket, not a plan!

The capability to assign start and due dates is a given. Here are some other key points to consider for scheduling, all of which are Premium features:

- **Dependencies:** Most simple plans won't have dependent tasks, but as project complexity grows, so does the need for dependencies and the capability to track them. Planner Basic has no dependency capability, and Planner Plan 1 adds Finish-to-Start dependencies. Planner and Project Plan 3 adds Start-to-Start, Start-to-Finish, and Finish-to-Finish dependencies.

- **Timeline view:** Planner Basic includes the Schedule view that enables you to view tasks on a calendar. This works well for smaller, simpler plans where you're not concerned about dependencies between tasks. When you do need to work with dependencies, however, Timeline view is a must. It enables you to visualize those dependencies and interrelationships.

- **Baselines:** A baseline captures a snapshot of a plan at a specific point in time. You use the baseline as a reference point to measure plan progress. Baselines are more useful in longer or more complex projects, so if you're building simple plans such as for self-improvement or learning a new skill, you won't need baselines.

- **Critical path:** The critical path in a plan is the longest sequence of dependent tasks that determine the earliest possible completion date for the plan. If one task on the critical path is delayed, the project is delayed. Simple plans, no. Complex plans, probably. Critical projects in an enterprise, absolutely. You get this capability with Planner and Project Plan 3 and Planner and Project Plan 5.

- **Milestones:** Project milestones are like checkpoints in your project. They indicate important dates, events, deliverables, or even decisions that are key to forward progress in your project. Unlike individual tasks that usually span time, a milestone has no duration. Tasks are the discrete actions that take you from point A to point B. Reaching point B is the milestone. You can use a one-day duration task in Planner to simulate a milestone, but true milestone capability comes with Planner Plan 1.

Collaboration and integration

The following features and capabilities enable you to collaborate with others on a project, immediately see their changes or additions, and use other tools in conjunction with your chosen planning tool:

- **Shared workspaces:** Does the solution give you the capability to share a workspace where you can store documents and other items that are important to the plan? Microsoft SharePoint is an example for Planner. Adding Planner to a Microsoft Teams channel gives you the capability to insert attachments from a SharePoint library in a task. If you add Planner in Teams without attaching it to a channel, you can only insert links in a task.

- **Real-time editing:** When you're collaborating with others on a project, you almost always need to see their updates in real time. All versions of Planner support real-time editing, but users who are offline won't see your changes until they reconnect. Notifications aren't real time either, but they require that you open the plan or receive an email notification.

- **Communication tools:** With Planner, communication mainly takes the form of chat in Teams. However, when you add a plan to a Teams channel, you gain the capability to have threaded conversations in Teams specifically about a task. Each task can have its own conversation thread. Any comments you add through the Planner web app, however, do not appear in the task's chat thread in Teams.

  Although the Planner web app is very handy for creating and managing plans, the benefits you get from creating a plan in a Teams channel are a strong argument for creating and working with them in Teams when you need to collaborate on a plan with others.

- **Notifications:** This isn't much of a requirement if you're only using Planner for yourself, but it becomes important when you need to assign tasks and collaborate with others on a plan. All versions of Planner provide the capability for notifications through email and Teams, although notification of tasks marked as complete is only available through Teams.

Resource management

There is only one resource for plans you create and execute yourself. You're it! But even in projects for small groups, the capability to identify and align resources, whether people or otherwise, can be important. Resource management becomes a must-have for enterprise-level projects.

Following are the key topics to consider for resource management when choosing a planning tool:

>> **Resource allocation:** Resource allocation can be as simple as assigning a task to someone on your team. It's a capability in all versions of Planner. All versions also give you ways to view who owns each task. Resource allocation can also be much more complex, where you need to manage consultants, engineers, or other skilled resources and ensure that they have availability for their assigned body of work. Planner and Project Plan 3 introduces higher-level resource-management capabilities. Planner and Project Plan 5 builds on those capabilities to add a full range of enterprise-wide resource management and allocation.

>> **Skills-based assignment:** Do you need a planning tool that lets you view skills held by your resources? For example, you wouldn't assign a framing carpenter to a tiling task or a software developer to a network configuration task. Planner Basic and Planner Plan 1 don't provide this capability natively, but you can implement some manual workarounds. Planner and Project Plan 3 gives you the capability to view resource skills and proficiencies. Planner and Project Plan 5 adds the capability to manage skills and proficiencies.

>> **Capacity planning and workload balancing:** This topic refers to the capability to plan and manage resource allocation based on resource capacity. Planner Basic enables you to assign resources to tasks but offers no People view or other features for viewing and managing resource capacity. Planner Plan 1 adds some rudimentary task-level features for viewing workloads across the team but provides no enterprise resource pooling. Planner and Project Plan 3 provides several features for capacity mapping and reporting, and Planner and Project Plan 5 provides portfolio-level capacity analytics for enterprise organizations.

Reporting

Reporting on a plan's progress quickly becomes important as project complexity and costs grow. Even a small project that you manage with Planner Basic can benefit from good reporting.

Following are some reporting topics to consider when choosing a planning solution:

>> **Progress tracking:** At a minimum, even with small team projects, you need a means of tracking how the project is progressing. Planner Basic provides limited built-in capabilities with the Charts and Schedule views, and you can expand beyond those capabilities by exporting data to other tools. Planner Plan 1 adds the Timeline view and goals reporting. Planner and Project Plan 3

adds reporting for critical path, portfolios, and road maps. Planner and Project Plan 5 adds a full suite of enterprise financial and resource management reports. Both Plan 3 and Plan 5 include native Microsoft Power BI support, whereas Planner and Planner Plan 1 require that you export data to use with Power BI.

>> **Built-in reports and dashboards:** You won't want to create all of your reports and dashboards manually regardless of which planning tool you choose. All Planner versions include built-in reports and dashboards, although those in Planner Basic are, well, basic. Think about the types of reporting that you need up front based on your planning needs, as well as what reporting or dashboards you'll need to create to supplement them.

>> **Customization capabilities:** Customization runs the gamut, from simple things like changing colors in the interface or adding custom fields to tasks, to using tools like Microsoft Power Automate to customize and automate the way the planning tool works. Surprisingly, as I write this, only Planner Basic fully supports Power Automate. This will change over time, but currently the Premium versions require Microsoft Project application programming interfaces (APIs) or Microsoft Graph API, Project Online connector, or Microsoft Dataverse.

>> **Integration with other reporting tools:** Do you use certain reporting tools now? If so, consider whether the planning tool you choose can integrate with those reporting tools, either natively or by exporting data.

Security, governance, and compliance

Wow! This one is a big and complex topic, so much so that I can't really do all of it justice within the context of this book. These topics require input from your security team, IT, legal counsel, and the other groups that manage these areas in your organization.

Even so, here are some key things to think about when choosing a planning solution:

>> **Authentication and authorization:** Your planning solution needs to control who can access it and what actions they can perform. At a minimum, this needs to be enforced at the plan level, but in more complex situations you may need more granular control.

Authentication identifies an individual, and *authorization* controls what they can do. For example, having a passport identifies you (authentication), while a visa enables you to visit a specific country (authorization).

>> **Data storage:** How your data is physically stored and protected is important to guard against unauthorized access, and *where* it is stored ensures data residency requirements. For example, your organization may require that your data reside within a specific geographic area for compliance reasons.

>> **Auditing:** In this context, *auditing* means tracking changes that happen within your plan, including such things as task assignment, deletion of items, and other changes. Certainly not much of a concern if you're only working on personal plans, but auditing can be important even in smaller organizations. Consider what capabilities the solution you're considering offers for change tracking, auditing, and reporting.

>> **eDiscovery, legal hold, and data life cycle:** *eDiscovery* is the ability to identify, collect, and review electronically stored information (like emails, documents, files, and messages). *Legal hold* is the requirement to retain specific data that could be relevant to some legal action. The *data life cycle* is generally the length of time that you must retain data to meet legal and regulatory requirements.

> Again, not a concern for personal use, but these are definitely considerations for businesses. Work with your legal department or legal counsel to determine what requirements you have for retaining and discovering data in the event of legal action, and the requirements for retaining data.

>> **Data loss prevention and information protection:** Identify and ensure your requirements are met for protecting against data being improperly shared or exposed to unauthorized users.

>> **External collaboration:** If you need to collaborate on plans with others outside of your organization, ensure that the solution you choose meets your organization's requirements for protecting against unauthorized access and provides an appropriate means for inviting and managing external users.

Some of the topics briefly described in this section are covered in more detail in Chapter 16.

Surveying Your Options

This book clearly focuses on Planner and Planner Plan 1 and offers some insights into the other premium versions of Planner. But it's also important to understand Planner's place in relation to some of the more popular planning solutions. This section takes a brief look at some of these solutions, starting with Project.

Microsoft Project

If you're not new to project planning and management, you're probably at least somewhat familiar with Project. Maybe you've used the Project desktop application, Project for the web, Project Server, or Project Online. Microsoft is redesigning, consolidating, and rebranding Project, so it's important that you understand those transitions.

Project for the web

This web-based planning solution from Microsoft offered features and capabilities similar to Planner and its Premium features. Project for the web has been retired as of August 2025 and rolled into Planner. Existing plans are redirected to the new Planner experience, and the Project and Roadmap tabs for Teams have been deprecated. The Premium features — like sprints, task dependencies, Gantt charts, and others — have moved into the various Planner Premium versions.

Project Online

You can think of Project Online as encompassing the features in Planner and Project Plan 3 and Planner and Project Plan 5. As its name implies, Project Online is an online project portfolio management (PPM) solution that could work in conjunction with or separately from Project for the web. As of October 2025, Project Online is no longer being sold. It's set for official retirement on September 30, 2026, at which time it will no longer be available.

Project Online remains available until that time, but if your organization is using it, you should start a transition plan soon and be ready to migrate any remaining projects from Project Online to Planner prior to the retirement date.

Project Server

Project Server is Microsoft's on-premises PPM solution providing a full range of project management features and capabilities. Many organizations choose Project Server because they have data residency or other requirements that require them to host all their data in-house. Others are still transitioning from a data-center model to a cloud model, so they continue to host project management in their data centers. Extended support for Project Server ends July 14, 2026. Project Server Subscription Edition will be supported through at least 2029.

The Project desktop app

Using the Project desktop app, you can create and manage plans locally on your own computer for single-project scheduling with a full range of PPM features, including task dependencies, critical path, baselines, progress control, and other features, all with multiple built-in views and reporting. When used as a front end to Project Server or Project Online, the desktop app becomes a scheduling front end and leverages the enterprise features for portfolio analysis, resource management, and other enterprise features in Project Server.

You can import data into Planner from other tools by copying and pasting, using Microsoft Excel as an intermediary, or using automated solutions like Power Automate. Which option you choose depends on the options you have for exporting the data from the other tools, the type of data, and other factors. Because these factors can vary quite a bit, I don't cover data migration here. Just keep in mind that if you're considering moving to Planner from some other project management tool, the chances are good that you'll be able to move some, if not all, of your data, into Planner. Do your research ahead of time to understand the possibilities.

Other tools

The earlier section, "Defining Your Needs," explores various features and capabilities of a planning solution in a general way but also provides context for Planner's versions. Because this chapter approaches PPM in a general way, it's also useful to explore a few of the more popular non-Microsoft planning solutions, because Microsoft has by no means cornered the PPM market. That doesn't mean that Planner is the wrong solution for you, but any reference should at least cover alternatives to help you validate the choice you've made, or at the very least, expand your knowledge base. This section does just that.

The market share numbers I include in the following sections aren't written in stone. Measuring market share can be an arcane art, and the results depend on who you ask. The numbers I cite are directionally correct for this discussion.

Jira

Atlassian's Jira holds just under 8 percent in the enterprise PPM software space compared to Microsoft's roughly 6 percent. Jira has a strong following in the software development and DevOps spaces. It offers a strong set of enterprise PPM features, scales to large organizations, and includes integration to GitHub and Bitbucket, which helps in part to explain its popularity for software development.

On the con side, the learning curve for nontechnical, non–project management office (PMO) teams can be steeper than other solutions like Planner. It's also a lot of solution for small plans. You probably wouldn't use Jira for personal projects!

Atlassian offers a free version for up to ten users, and its Standard and Premium versions are competitively priced. You have to contact Atlassian Sales to get a quote for their Enterprise version. For more information on Jira, visit www.atlassian.com/software/jira.

Trello

Trello is another planning offering from Atlassian, and the entry-level version is also free. Trello is super simple and easy to use, requiring little to no training. It's useful for light teamwork and personal task planning. The fact that it's free and so easy to use is, in part, why it boasts more than 90 million registered users.

There are a host of bolt-on features that you can add to Trello to gain additional capabilities, as well as three paid versions, each of which adds further capabilities.

You can find more about Trello at `https://trello.com`.

Chapter **4**

Stepping into Microsoft Planner

As more applications move to a web-based, software as a service (SaaS) model, user interactions with those applications most often occur through a web browser. Your interaction with Microsoft Planner takes that form but adds a second option: Microsoft Teams. Both options work well but offer trade-offs. The web interface may be slightly faster, but Teams gives you access to some features that you can't access from the web app. I explore these differences in Chapters 5 and 8. This chapter provides an overview of opening and accessing Planner in both scenarios and covers the basic interaction of creating a plan from a template, naming a plan, and managing individual plans.

Opening Microsoft Planner

You can open Planner in two ways: through a web browser or from Teams. I've worked with Planner in both the web application and Teams, and I don't have a preference. Other than a bit of functionality, the only difference is the presence of the app bar (also known as the navigation pane) on the left side of the Teams window where the Teams apps are pinned. Otherwise, you work with Planner in both apps in much the same way. I choose Teams when I need those features that are available only in Teams, and I choose the web app when I don't need that integration or I'm working on a personal plan.

Following is a list of the web browsers that are compatible with Planner, depending on your operating system:

>> **Windows 10 and Windows 11:** You can use the latest versions of Google Chrome, Microsoft Edge, and Mozilla Firefox.

>> **macOS:** You can use Safari version 16 or later, Edge, or Chrome.

>> **iOS:** You can use Safari version 16 or later, Edge, or Chrome, as well as the Microsoft Planner mobile app, which offers a better user experience than a browser.

>> **Linux:** You can use Edge, Chrome, and Firefox.

>> **Android devices:** There is currently no officially supported browser for Android, but you can use the Microsoft Planner mobile app to access Planner.

To open Planner in a web browser, just open a compatible browser and go to `https://planner.cloud.microsoft`. Figure 4-1 shows the results.

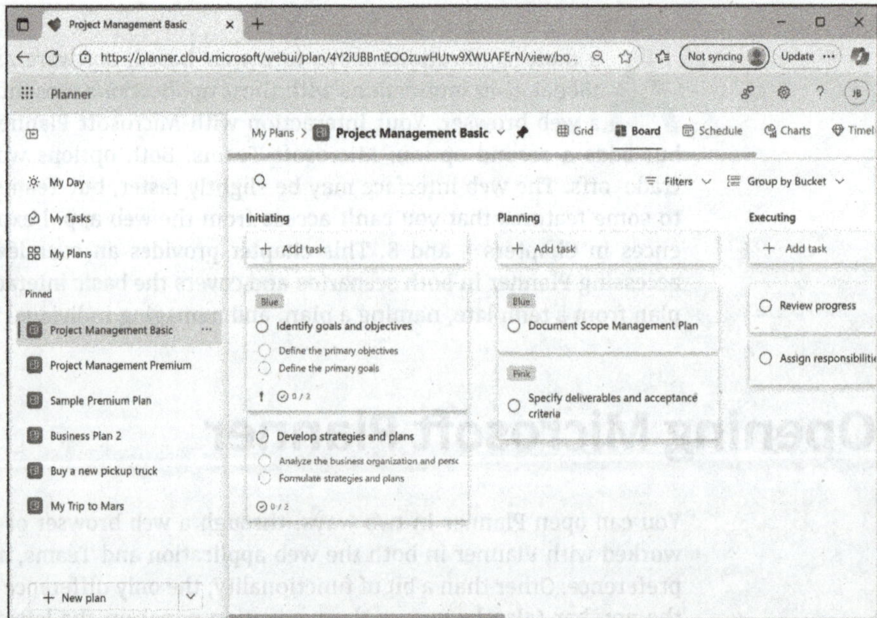

FIGURE 4-1:
You can use a web browser to work with Planner.

Opening Planner in Teams takes a few more steps the first time around. Follow these steps to pin Planner to the app bar and open Planner in Teams:

1. **Open Teams.**

2. **Click the Apps icon at the bottom of the App bar (see Figure 4-2).**

Apps icon

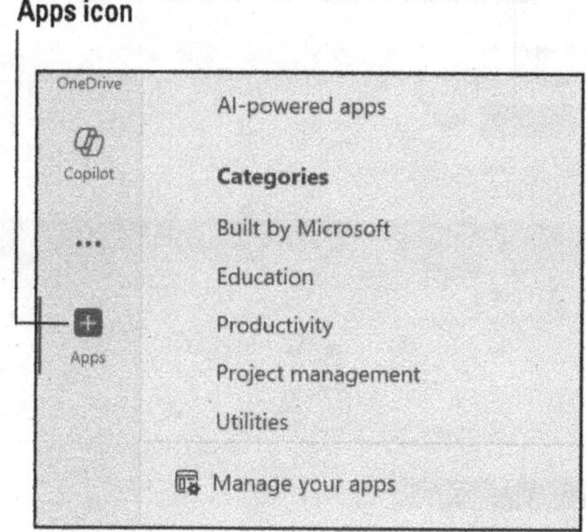

FIGURE 4-2:
Click the Apps
icon in the
Teams App bar.

3. **Click in the Search bar (see Figure 4-3) and type** Planner.

    The results of the search appear on the right side of the screen.

Search bar

Add button

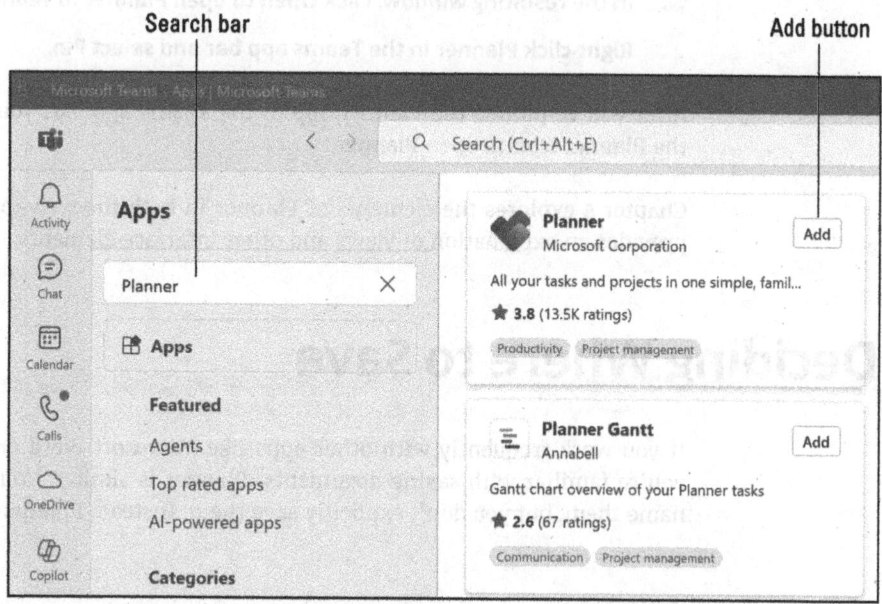

FIGURE 4-3:
Search for the
Planner app
to add it.

4. **Click the Add button next to Planner (refer to Figure 4-3).**

   The Planner description page appears (see Figure 4-4).

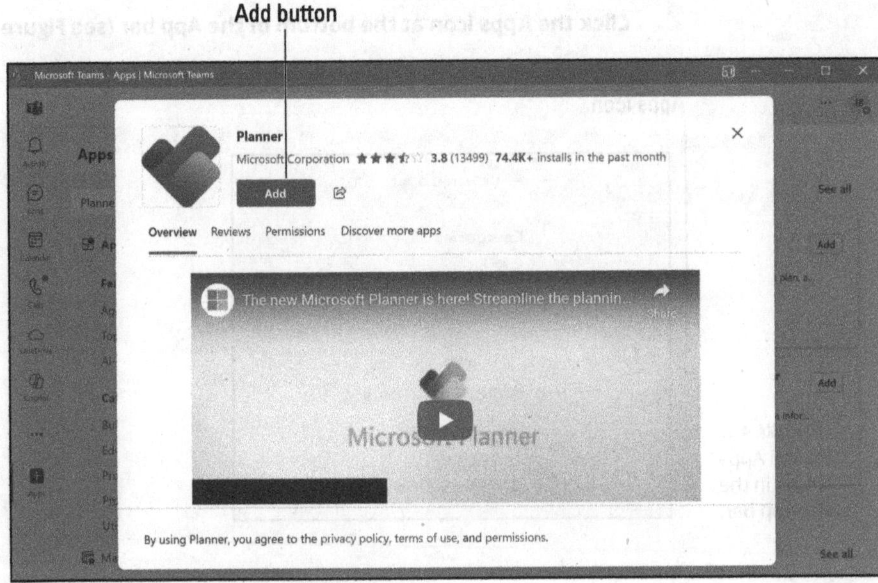

FIGURE 4-4:
Teams provides a description of the Planner app before you add it.

5. **On the Planner description page, click the Add button.**

6. **In the resulting window, click Open to open Planner in Teams.**

7. **Right-click Planner in the Teams app bar and select Pin.**

After you've pinned the Planner app to the Teams app bar, you can simply click the Planner icon to open Planner.

Chapter 5 explores the elements of Planner in both the web app and Teams and provides an explanation of views and other interface elements.

Deciding Where to Save

If you work frequently with other apps like Microsoft Word or Microsoft Excel, you're familiar with saving documents. Planner is similar: You create plans and name them, but you don't explicitly save them. Instead, Planner takes care of that for you automatically. The question is, "Where?" Unlike those other apps, it's not in a folder on your local storage or in Microsoft OneDrive. The answer also depends in part on whether you're using a Teams-integrated plan.

For personal plans that are not associated with a Microsoft 365 group, Planner stores task metadata in Microsoft Azure. This metadata includes plan names, task titles, task descriptions, and other task properties. Comments are stored in your Microsoft Exchange Online mailbox. Attachments associated with personal plans are stored in your OneDrive.

The term *metadata* refers to properties that are applied to data. For example, a Word document includes information about the document's author, the date it was created, and the file size, among other properties. An Excel document has similar metadata attached to it. Metadata can be used in many ways, including making it easier to search for and find specific documents and enabling systems to automatically identify documents and data types.

There is no native backup solution for personal plans, so after you've deleted a plan, you won't be able to get it back. You can, however, use a third-party backup solution to back up your plans and their related data. The one I'm most familiar with is AvePoint Cloud Backup, but there are several backup solutions that work for Planner. If you need to add backup and restore capability for your personal plans (or gain additional features for your Microsoft 365 Groups–enabled plans), use your favorite search engine or artificial intelligence (AI) to search for "Microsoft Planner Basic backup solutions."

Teams-integrated plans take advantage of Microsoft 365 Groups and the services associated with them. Like personal plans, Microsoft 365–linked plans' metadata (task titles, descriptions, dates, and so on) is stored in Azure. Comments are stored in the Microsoft 365 group's Exchange Online mailbox, and attachments are stored in the group's Microsoft SharePoint Online document library. Storing Planner data in these Microsoft 365 resources enables integration with Microsoft and third-party solutions for eDiscovery, retention, and compliance. However, even Microsoft 365–linked plans are not easily recoverable without a third-party solution. Instead, a Microsoft 365 admin must restore the entire group, which includes the group's SharePoint site and group mailboxes and files. It's an all-or-nothing proposition.

To sum it up, you don't need to save your plans explicitly because Planner does it for you. But you do need to be careful not to delete your plans unless you're sure you no longer need them or you have a backup and restore solution already in place and have tested it fully to make sure it works as you expect it to.

Creating a New Plan

Creating a new plan in Planner is easy. You start by clicking New Plan in the Planner navigation pane. But at that point, you need to make a decision.

Deciding between Basic and Premium

Clicking New Plan opens the Create New window, shown in Figure 4-5. Here you make the choice between a "clean" Basic or Premium plan, or choose a Basic or Premium plan created from a template. A clean plan, whether Basic or Premium, contains no tasks, buckets, or customizations — it's a clean slate. You can then add tasks, buckets, and other customizations as needed.

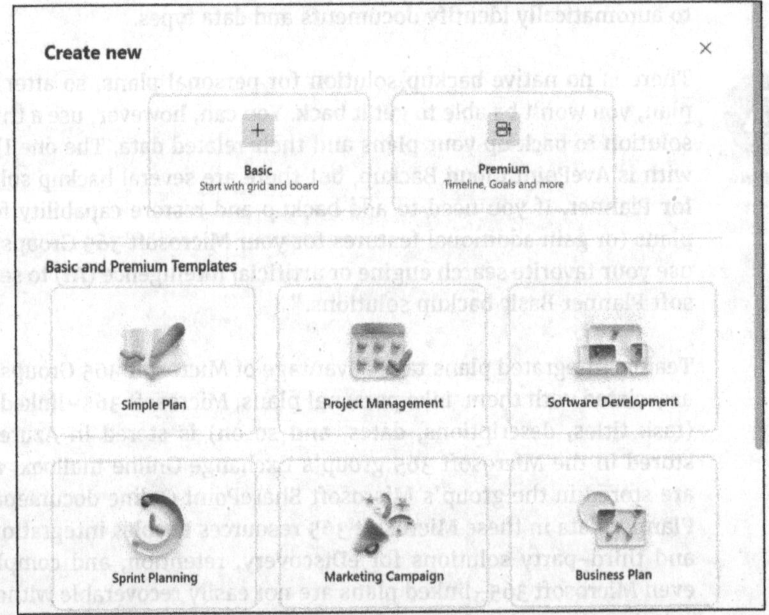

FIGURE 4-5: Use the Create New window to create a clean plan or choose a template.

If you don't have a Premium license, your decision is made for you. You can only create a Basic plan. Even so, you're not limited to the clean Basic plan. You can choose from a small selection of Basic plan templates, just as you can for a Premium plan.

Choosing a template

The Create New window (refer to Figure 4-5) offers a selection of templates in addition to the clean Basic and Premium plan options. Scroll down in the window and choose See All Templates to view all available templates. The Templates window (shown in Figure 4-6) gives you access to all of them. Click a template in the left pane to view information about the template in the right, including whether it's a Basic plan or Premium plan. When you find the one you want, just click Use Template to create the plan using the selected template.

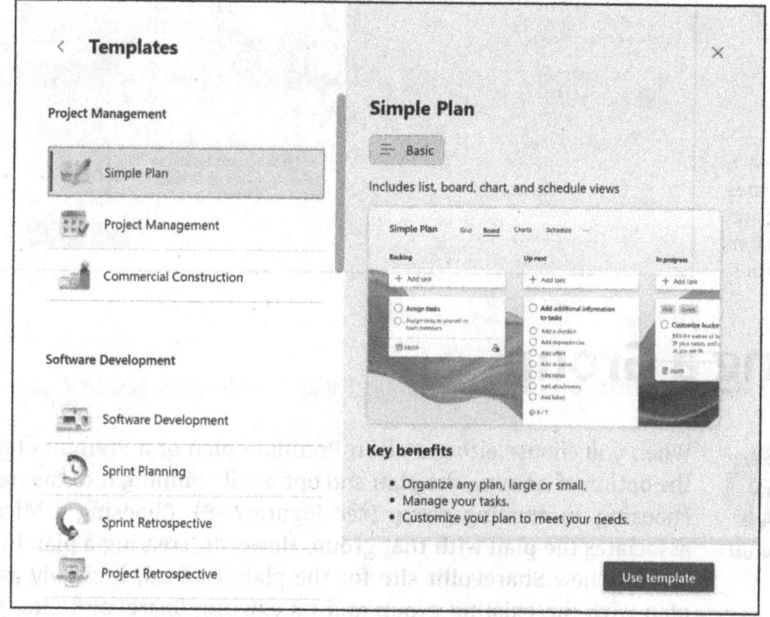

FIGURE 4-6: View information about all plans in the Templates window.

Some templates let you choose between Basic and Premium versions. The Project Management template, for example, offers both options as shown in Figure 4-7. The Basic option includes preconfigured buckets and other elements that take care of some of the common setup actions for you. The Premium option adds a Timeline view and other Premium features.

You may be wondering if you can create your own templates or acquire templates other than the ones offered in the Templates window. Yes! I cover that topic in Chapter 11.

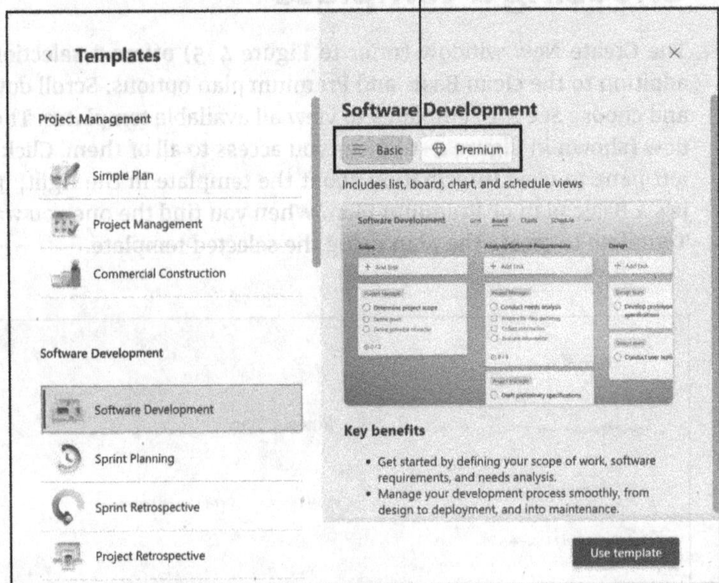

FIGURE 4-7:
Some templates offer Basic and Premium versions.

Adding a Group

PREMIUM

When you choose either a clean Premium plan or a Premium template, you have the option of naming the plan and optionally pinning it to the navigation pane, or choosing an existing group (see Figure 4-8). Choosing a Microsoft 365 group associates the plan with that group. However, creating a plan in this way doesn't create a new SharePoint site for the plan. Instead, it simply associates the new plan with the existing group and its existing SharePoint site, group mailbox in Exchange Online, and Teams channel.

FIGURE 4-8:
You can choose an existing Microsoft 365 group when creating a Premium plan.

If your account is a member of a group that has permissions to create Microsoft 365 groups, you'll also have the option to create a new group when creating a plan. Using this option creates the Microsoft 365 group along with its corresponding SharePoint site and other group resources.

Turn to Chapter 16 to find out more about managing Microsoft 365 group creation permissions for Planner.

Naming (and Renaming) Your Plan

If you've stepped through the process of creating a plan, you already know how to name the plan. Just type the name in the Name text box when creating the plan. Renaming a plan isn't as obvious. Here's how to do it:

1. **In Planner, open the plan you want to rename.**

2. **In the Work pane, click the drop-down arrow next to the current plan name (see Figure 4-9) and choose Plan Details.**

   The Plan Details window, shown in Figure 4-10, appears.

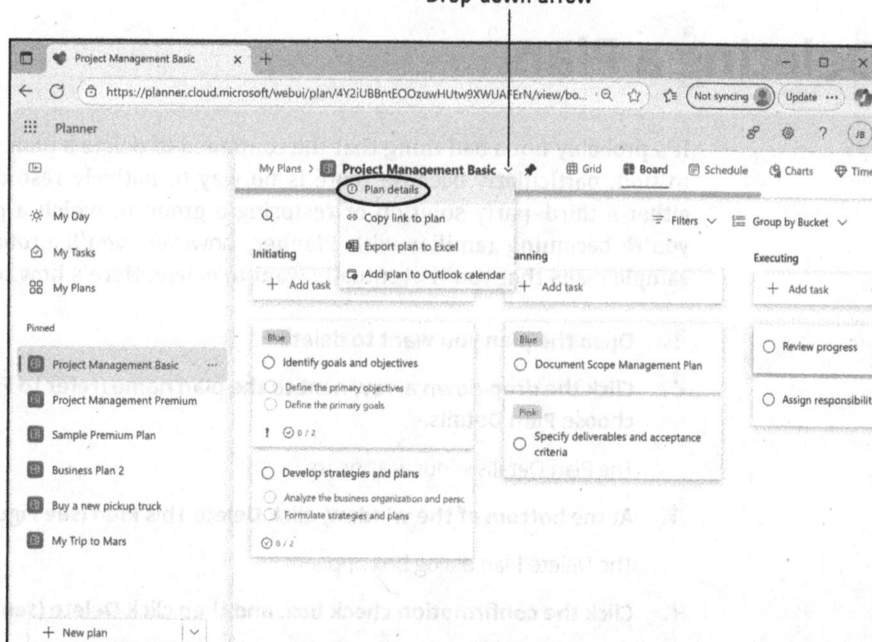

FIGURE 4-9: Choose the drop-down to view the Plan details window.

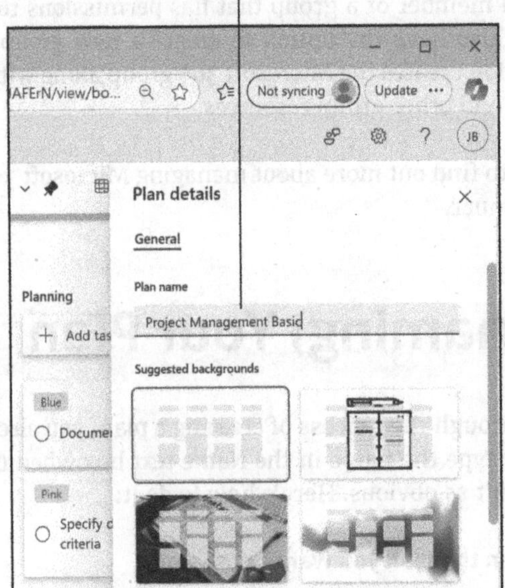

FIGURE 4-10:
Use the Plan Details window to rename a plan.

3. **Type a new name in the Plan Name text box.**

4. **Close the Plan Details window.**

Deleting a Plan

It's probably not a bad thing that the command to delete a plan takes a few clicks to find, particularly because there is no way to natively restore a plan without either a third-party solution or restoring a group to which a plan is linked. As you're becoming familiar with Planner, however, you'll probably create some sample plans that you'll eventually want to delete. Here's how to delete a plan:

1. **Open the plan you want to delete.**

2. **Click the drop-down arrow next to the plan name (refer to Figure 4-9) and choose Plan Details.**

   The Plan Details window appears.

3. **At the bottom of the window, click Delete This Plan (see Figure 4-11).**

   The Delete Plan dialog box appears.

4. **Click the confirmation check box, and then click Delete (see Figure 4-12).**

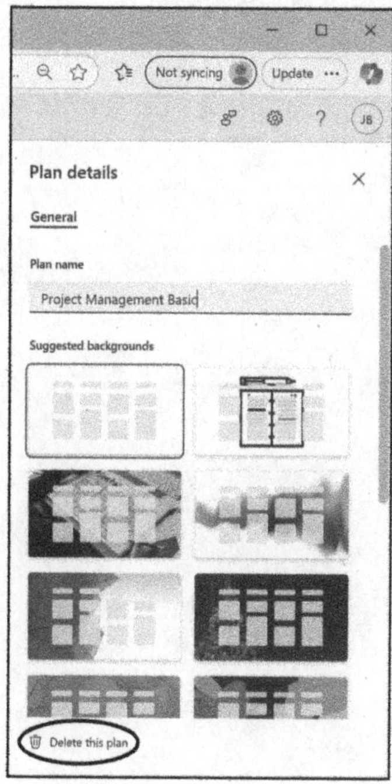

FIGURE 4-11: The option to delete a plan is at the bottom of the Plan Details window.

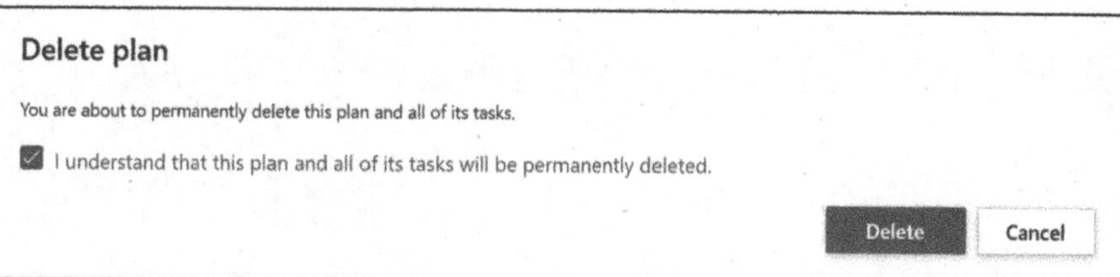

FIGURE 4-12: Confirm the deletion by checking the box and then click Delete.

2
Grasping the Basics of Microsoft Planner

Chapter **5**

Navigating the Microsoft Planner Interface

Poking around in an application's interface is a good way to learn how to work with the application. But it's also a great way to learn what the application can do, provided you have a guide along the way. That's what this chapter is all about, and a good place to start is the navigation pane and work pane.

Navigating the Planner Interface

Whether you're using Microsoft Planner from a web browser or from Microsoft Teams, Planner looks and functions mostly the same. I say "mostly" because there are a few capabilities in Teams that you can't get in the web versions. I highlight these capabilities throughout this book whenever appropriate. At a basic level, however, Planner has two main interface elements: the navigation pane and the work pane.

Using the navigation pane

Figure 5-1 shows the Planner web interface with the navigation pane and the work pane highlighted. The navigation pane on the left enables you to quickly open the My Plans, My Tasks, and My Day views (explored in the following sections). It also includes a Pinned section where you can pin and quickly open frequently used plans.

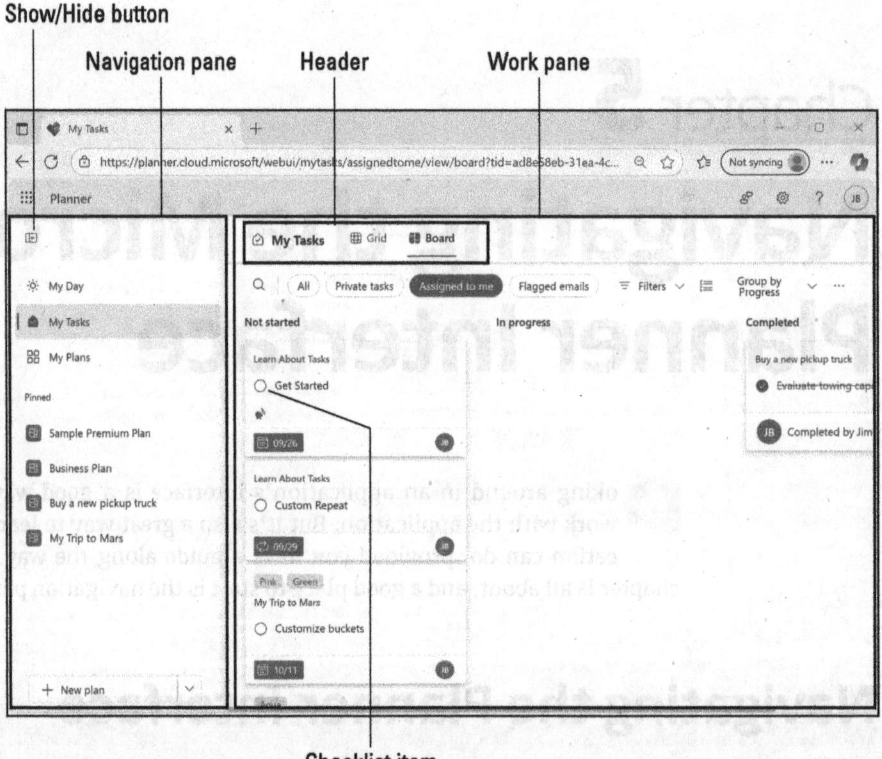

FIGURE 5-1: The navigation pane and the work pane are the key elements of the Planner interface.

The work pane on the right is where you'll work with Planner objects like tasks and views. Sometimes you'll work with those objects directly in the work pane. For example, when a task includes a checklist, you can click an item in the list to mark it as complete. You don't have to open the task to do so. In Grid view, you can click the name of a task to rename it without opening the task in a separate window.

All of Planner's views are explained in detail in the "Exploring Views" section, later in this chapter.

TIP

The navigation pane is great for quickly switching views or opening plans, but when you're deep into managing a particular plan, it just takes space away from the work pane. Fortunately, you can hide the navigation pane when you don't need it. As called out in Figure 5-1, there is a Show/Hide button at the upper left of the navigation pane that you can use to show or hide the pane. When the navigation pane is open, just click the Hide button to hide it and gain more space for the work pane; Figure 5-2 shows the results. To show it again, click the Show button (see Figure 5-3).

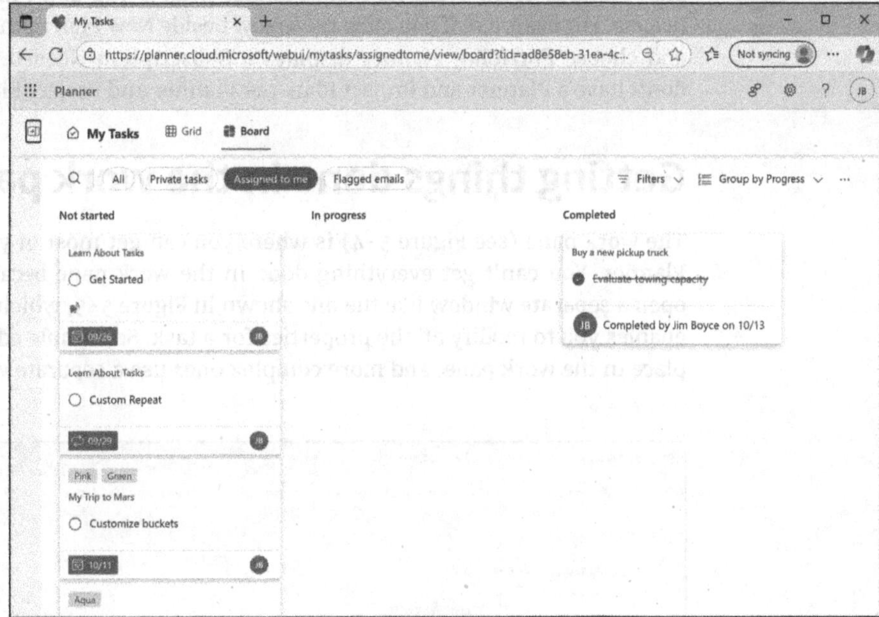

FIGURE 5-2:
Hide the navigation pane to give more space to the work pane.

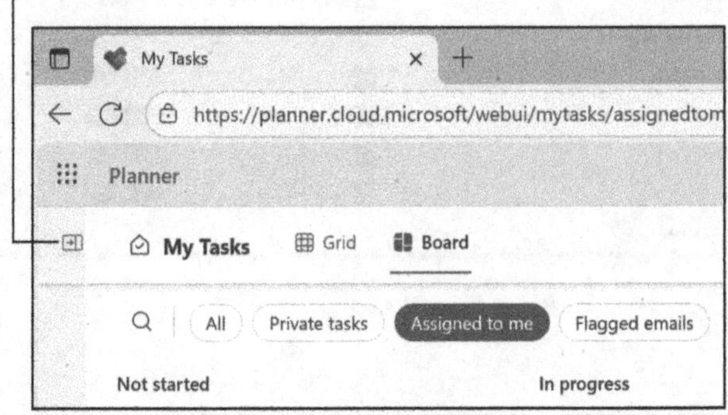

FIGURE 5-3:
Click the Show button to restore the navigation pane.

At the bottom of the navigation pane, you'll find a control that, by default, is labeled New Plan. Clicking the control opens the Create New dialog box, where you can choose between creating a blank Basic or Premium plan or choosing a template for the plan. If you have a Planner and Project Plan 3 or Planner and Project Plan 5 license, you can click the arrow to the right of the control to create a new portfolio of plans.

With Planner Basic you can create only Basic plans. You need one of the Premium plans, including Planner Plan 1, to create a Premium plan. Also, some Premium plan features seem available from Planner even if you don't have the necessary license. For example, if you click the arrow beside New Plan in the navigation pane and choose Portfolio, Planner displays a Subscription Needed dialog box if you don't have a Planner and Project Plan 3 or Planner and Project Plan 5 subscription.

Getting things done in the work pane

The work pane (see Figure 5-4) is where you can get most of your work done in Planner. You can't get everything done in the work pane because some actions open a separate window like the one shown in Figure 5-5, which, in this example, enables you to modify all the properties for a task. So, simple editing actions take place in the work pane, and more complex ones use a separate window.

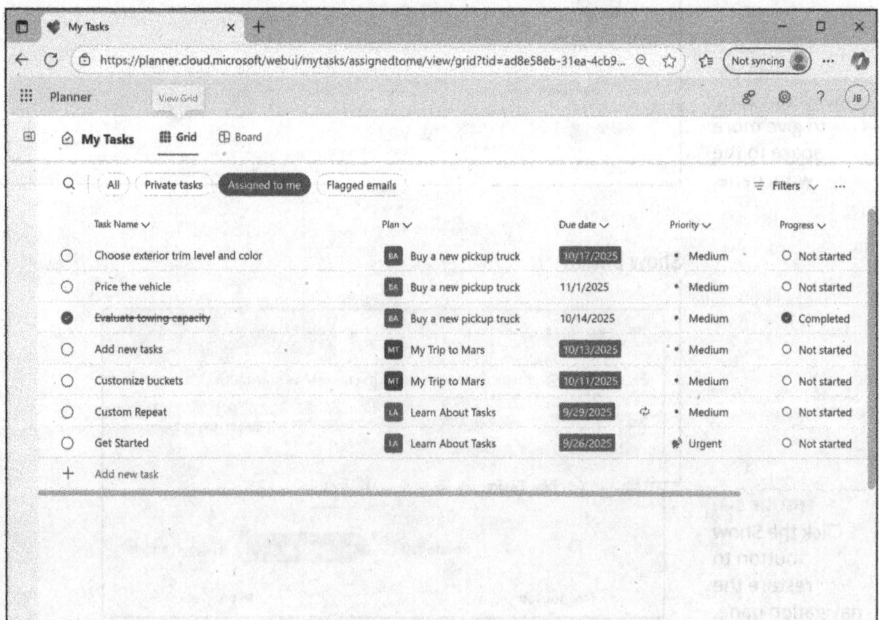

FIGURE 5-4: You can accomplish most of your work in the work pane.

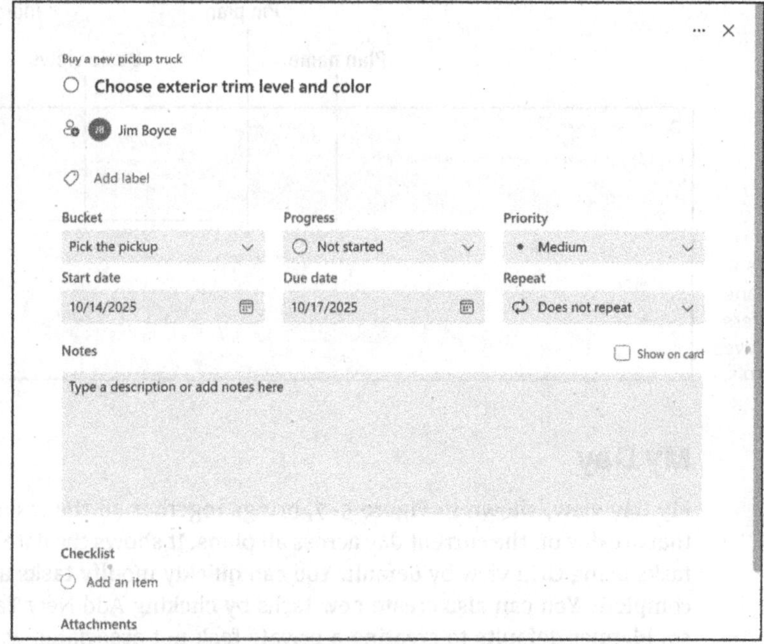

FIGURE 5-5:
Planner uses separate windows for more complex activities.

Across the top of the work pane is a header that includes a few controls for manipulating the pane. The first is a title that indicates the view you're using; in this example, it's My Tasks view. The header also lets you choose a specific view, such as Grid, Board, Schedule, Charts, or Timeline.

TIP

"Hold on!" you say. "My Tasks is a view, but so are Grid and Board?" That's right. My Day, My Tasks, and My Plans are all views. Grid, Board, Schedule, and other views give you different ways to see and manipulate the contents of those views. Just think of all these views as ways to organize, filter, sort, and work with Planner items.

The controls in the header change based on the current context. If you click a plan in the navigation pane, for example, the header controls change to give you access to features for working with the plan. These include controls to manipulate the plan (copy, export, and view details), choose a view, view task assignment and status, add team members, and other actions. Figure 5-6 shows an example with callouts.

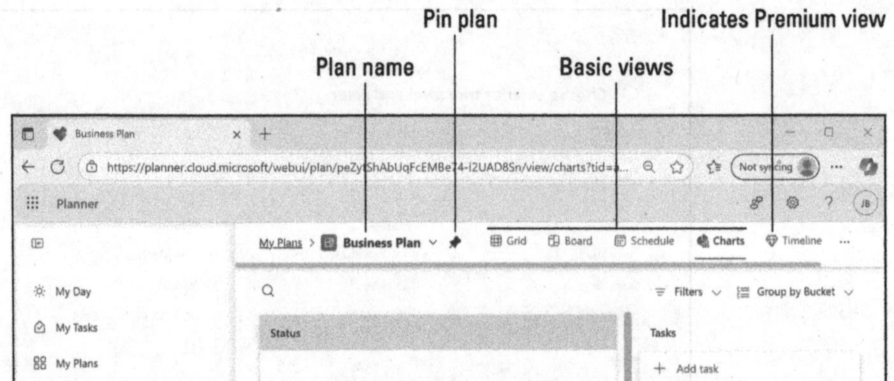

FIGURE 5-6:
The work pane header offers context-sensitive controls.

My Day

My Day view, shown in Figure 5-7, brings together all the tasks assigned to you that are due on the current day across all plans. It shows the date and lists relevant tasks using Grid view by default. You can quickly modify tasks and mark them as complete. You can also create new tasks by clicking Add New Task. When you do so, Planner defaults to creating a private task not associated with any plan. This gives you a way to add tasks that you need to complete, whether personal or work-related. You can view these tasks in My Day or click the Private tasks button in My Tasks view to list only your private tasks.

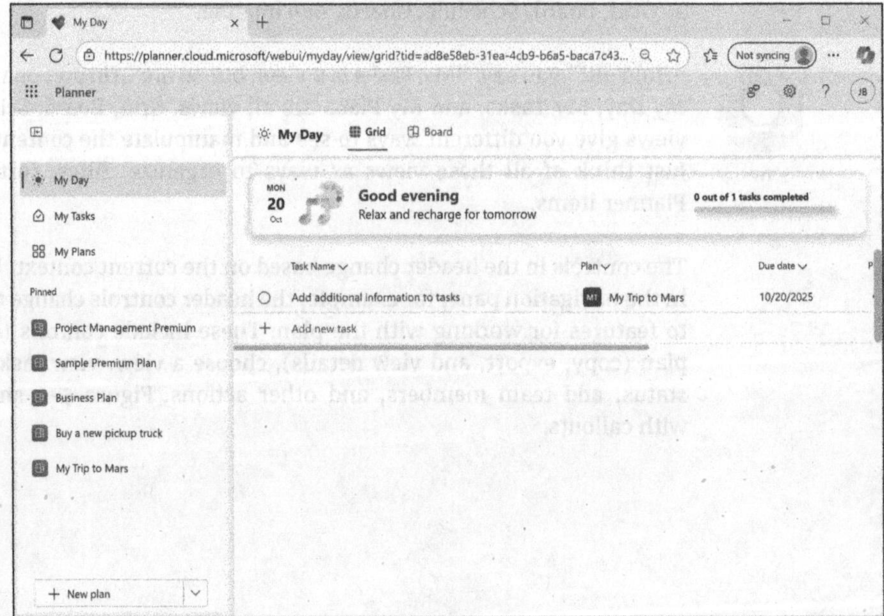

FIGURE 5-7:
My Day brings together all tasks assigned to you across all plans.

Even though you can create a personal task from My Day view, doing so doesn't assign the current day as the due date. You must add a due date if you want the task to be due on a specific date.

My Tasks

My Tasks view, shown in Figure 5-8, brings together all your tasks in one place. This includes tasks specifically assigned to you, as well as private tasks that you've created for yourself. As with My Day view, My Tasks shows all tasks regardless of their associated plan (or in the case of private tasks, no plan). You can use the view controls in the header to choose a view for your tasks, use the filters and grouping controls to filter and organize them. And edit tasks right in the view.

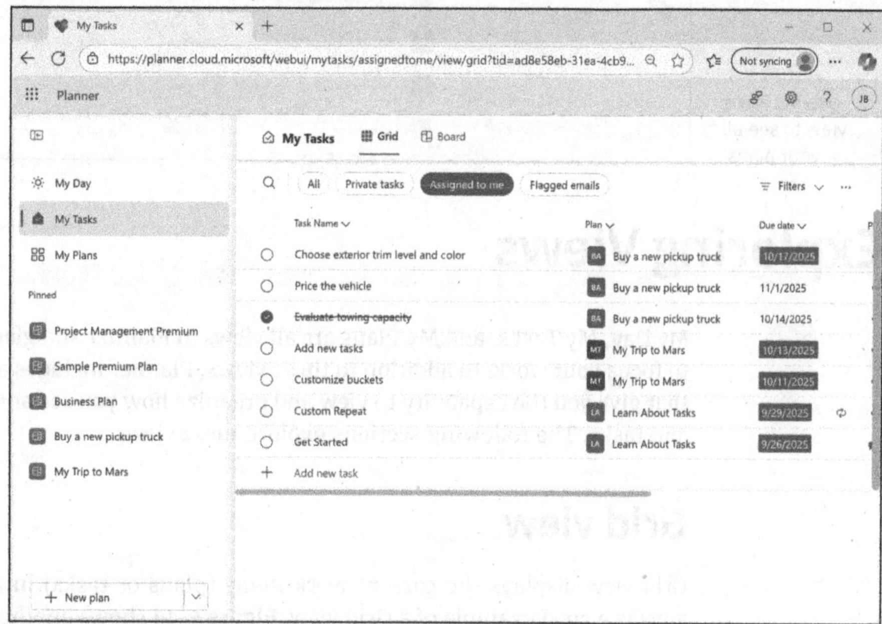

FIGURE 5-8: My Tasks view brings all your tasks into one place.

My Plans

My Plans view, shown in Figure 5-9, brings together all your plans in one place. It's great for quickly accessing your plans, particularly those that aren't pinned to the navigation pane. You can use the buttons in the header to choose the type of plans you want to view, like Shared or Personal. So, My Plans view is a great jumping-off point for all your plans. You can also click New Plan to create a new plan.

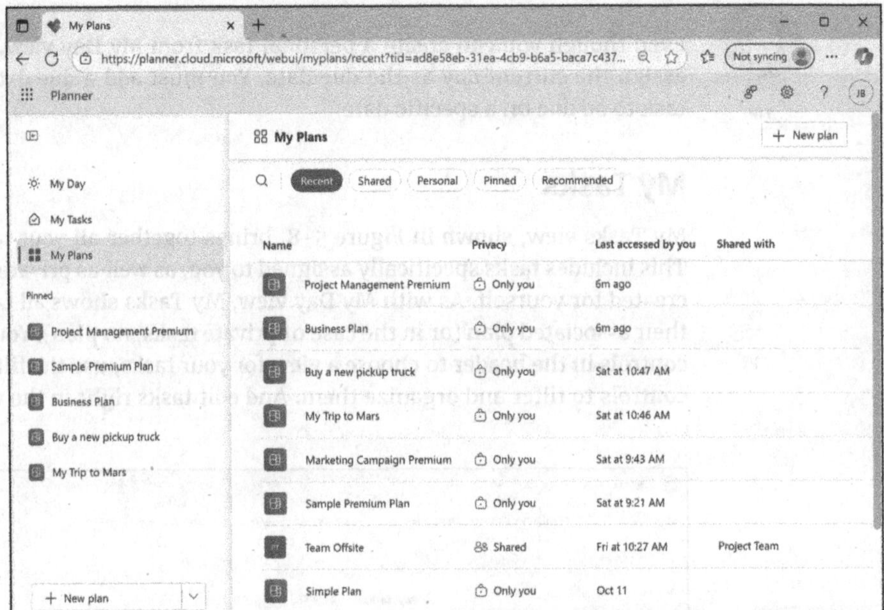

FIGURE 5-9: Use My Plans view to see all your plans.

Exploring Views

My Day, My Tasks, and My Plans are all views in Planner and give you an easy way to focus your work. In addition to these views, Planner includes a handful of views that give you the capability to view and organize how you see and work with plans and tasks. The following sections explore these views.

Grid view

Grid view displays the current work items (plans or tasks) in a table. My Plans view is a good example of a Grid view. Figure 5-10 shows another example: a plan showing all tasks in a Grid view. One advantage to using Grid view is the capability for in-place editing — you can edit items in the grid without opening them in a separate window. This makes it easy to edit multiple items without switching views. As with other views, you can easily filter the view to focus on specific types of plans or tasks.

Grid view shows all available columns by default. You can easily move columns around in Grid view to suit your needs. For example, maybe you want to view the Bucket column to the left of the Start Date column. Changing order is easy — just drag a column header into the desired position.

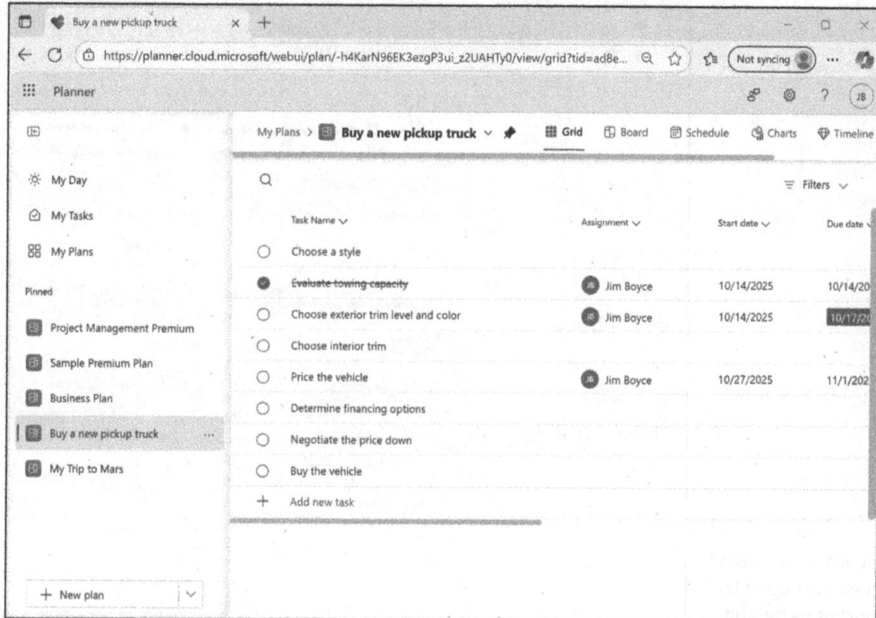

FIGURE 5-10:
Use Grid view
to show items
in a table.

PREMIUM

Although you can change column order in Planner Basic, you can't hide or unhide columns. You can do so if you have a Premium license but only if you're working in a Premium plan. Here's how to hide and unhide a column in a Premium plan:

1. **Click the column header and choose Hide Column.**

   The column immediately disappears from the grid.

2. **To unhide the column, scroll all the way to the right in the grid and click Add Column.**

3. **Click the column in the resulting column list.**

TIP

One odd behavior with Premium plans is the inability to sort a column. Clicking a column header in a Basic plan gives you the option to sort ascending or descending. The same action in a Premium plan only gives you the option to hide the column. I'm sure Microsoft will fix this eventually.

Board view

Board view offers a way to view tasks like sticky notes on the wall, grouped by bucket. As with Grid view, you can make some modifications to a task in Board view, but only in a very limited way, such as clicking a checklist. Clicking on a task in Board view opens the task in a separate window, where you can edit all the properties of the task. Figure 5-11 shows a Board view, and Figure 5-12 shows a task detail window.

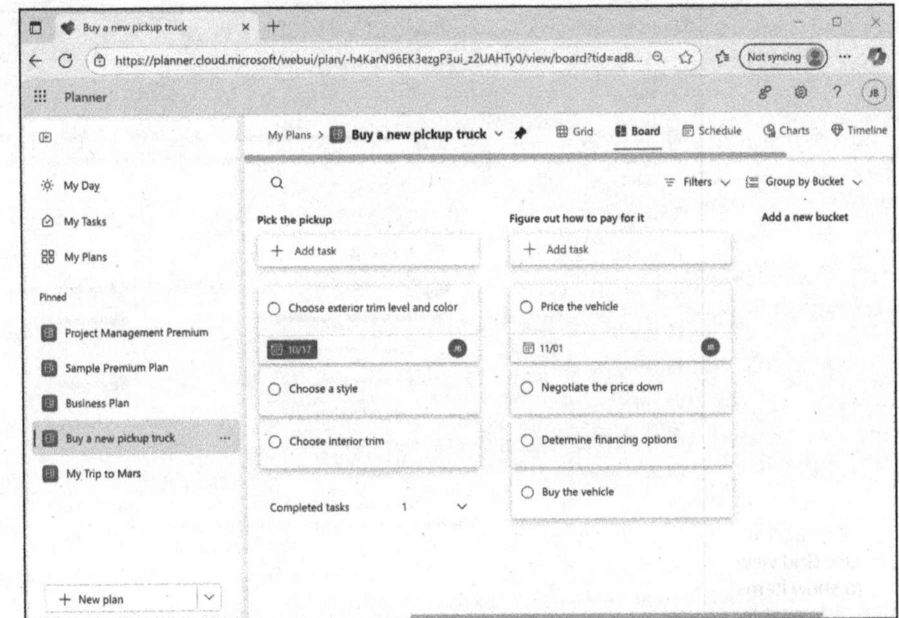

FIGURE 5-11:
Board view shows tasks grouped by bucket by default.

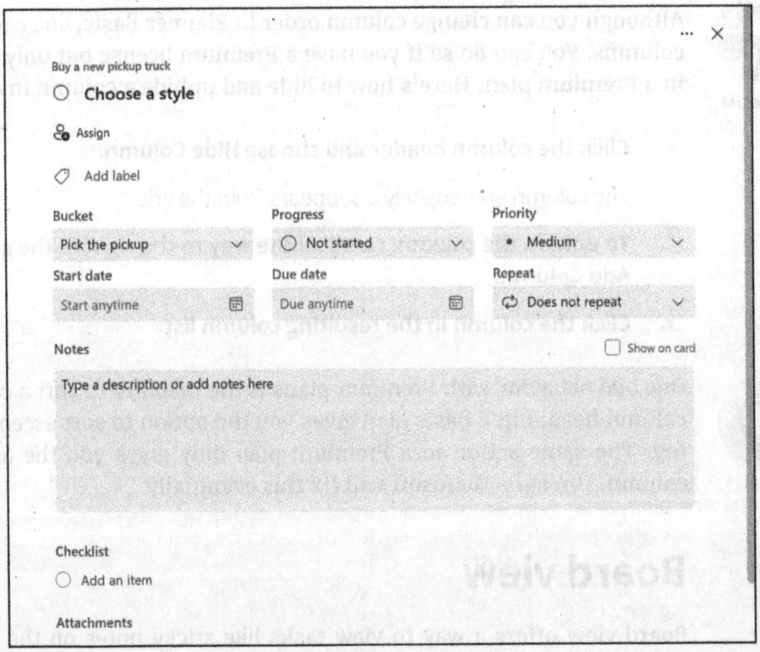

FIGURE 5-12:
You can use the task details window to modify all task properties.

By default, Board view groups tasks by bucket. You can easily change the grouping by clicking the Group by Bucket drop-down list at the top of the view and choosing a property like Assigned To or Progress. You can also limit the amount of information shown by using the Filters drop-down menu. For example, if you only want to view late tasks, click Filters, choose Due Date, and then choose Late, as shown in Figure 5-13.

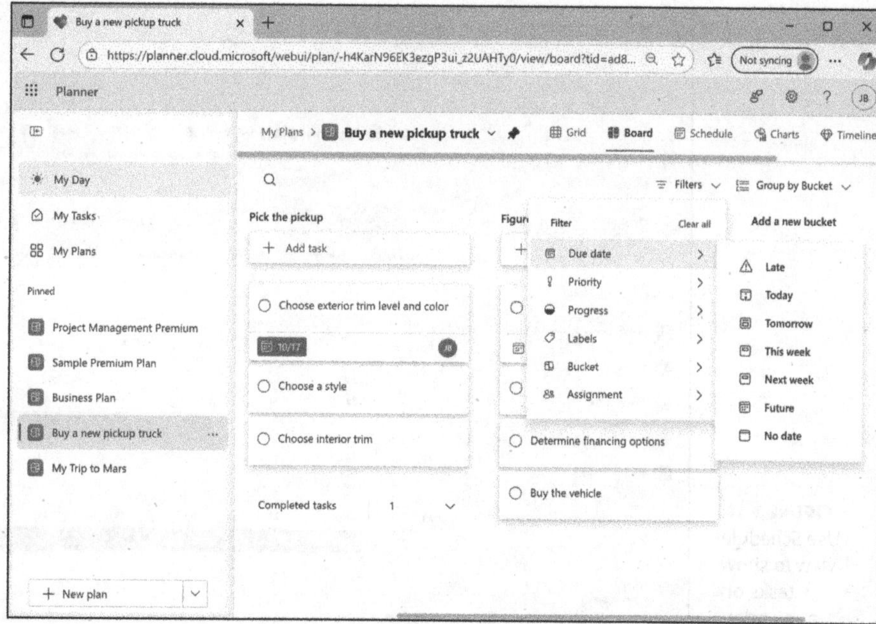

FIGURE 5-13:
Use the Filters drop-down menu to filter the view.

TECHNICAL STUFF

You'll hear boards in project portfolio management (PPM) tools referred to as *kanban boards.* The name originated in the 1940s as an element of the Toyota Production System, which sought to improve manufacturing efficiency. I don't speak Japanese, but Microsoft Copilot tells me that *kanban* translates to "signboard" or "billboard." Toyota used physical cards to signal various steps in the production process.

Schedule view

Schedule view shows tasks on a calendar with the duration of a task represented by the length of the task based on start and due dates (see Figure 5-14). The view is handy for smaller plans where you're more interested in when a task is due than how it relates to or is dependent on other tasks. Planner uses color to indicate when a task is active (gray), overdue (red), or complete (blue).

You can't change the colors used to indicate status in Schedule view, at least as I write this. This could be a feature that Microsoft adds in the future. A possible workaround is to use label colors in Board view to visualize status. Another option is to view the schedule in Microsoft Outlook, which offers more flexibility for conditional formatting using colors. See Chapter 8 for details on using Planner in conjunction with Outlook.

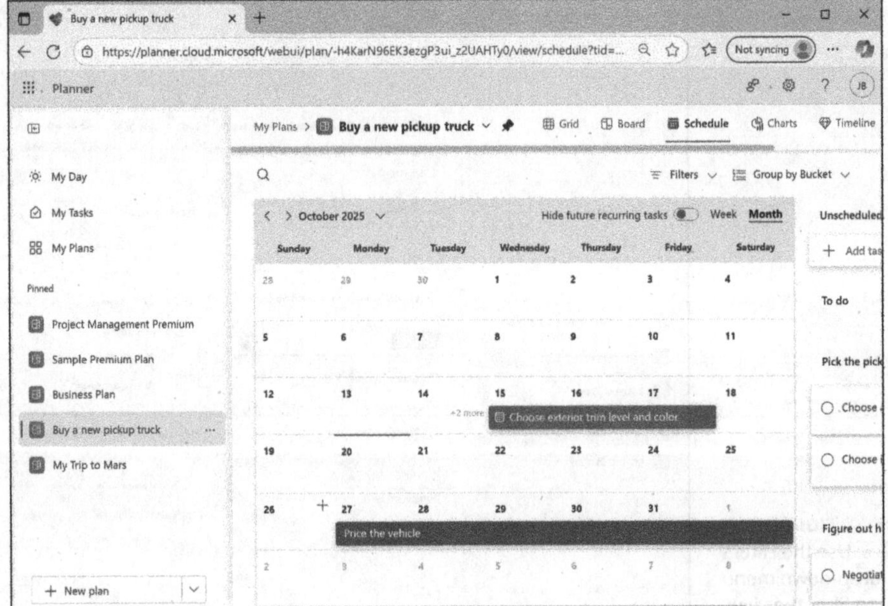

FIGURE 5-14: Use Schedule view to show tasks on a calendar.

Schedule view is somewhat limited in that it doesn't offer enough space to show the names of multiple tasks that are due on a given day. Instead, it displays a note on the day indicating the number of tasks, as highlighted in Figure 5-14. For Premium plans, a better option for more complex plans is Timeline view, which I cover later in this chapter.

Charts view

Charts view, shown in Figure 5-15, brings together a small number of visuals to show number of tasks by status, by bucket, and by priority. It also shows task number by team member assignment. As I write this, Charts view for a Basic plan offers more visuals than it does for Premium plans. Premium plans show number of tasks by status, number of tasks by bucket, and effort per person. The latter uses the number of hours for each task to determine total effort in hours.

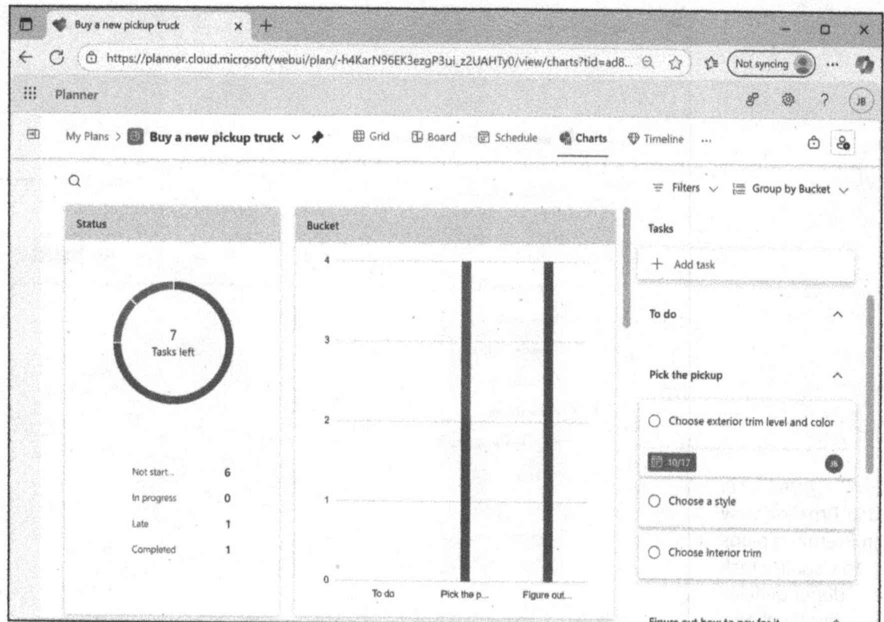

FIGURE 5-15: Charts view offers a small number of visuals to help you track plan progress.

TIP

Charts are not the only means for viewing status in Planner when working with a Premium plan. Premium views — including People, Goals, and Timeline — can be very useful for understanding resource allocation, task dependencies and status, and other indicators of progress, risks, and status overall.

Timeline view

PREMIUM

Timeline view (see Figure 5-16) offers a wealth of information about your plan using a Gantt chart model. A Gantt chart provides a way to visualize task start and due dates across a timeline, shows task dependencies, helps identify critical path items, and overall provides a great means of understanding the current state of your plan. Timeline view is indispensable for complex plans.

Like other Planner views, Timeline isn't just a static representation of current plan status. The task list at the left not only serves to identify tasks but also lets you mark them as completed. Tasks change color in the view when you mark them complete, further giving you a means of viewing task status in the view. Filtering is also available in the view. Click Filters to open a pane where you can choose a variety of filter criteria, including finish date, progress status, labels, buckets, and more.

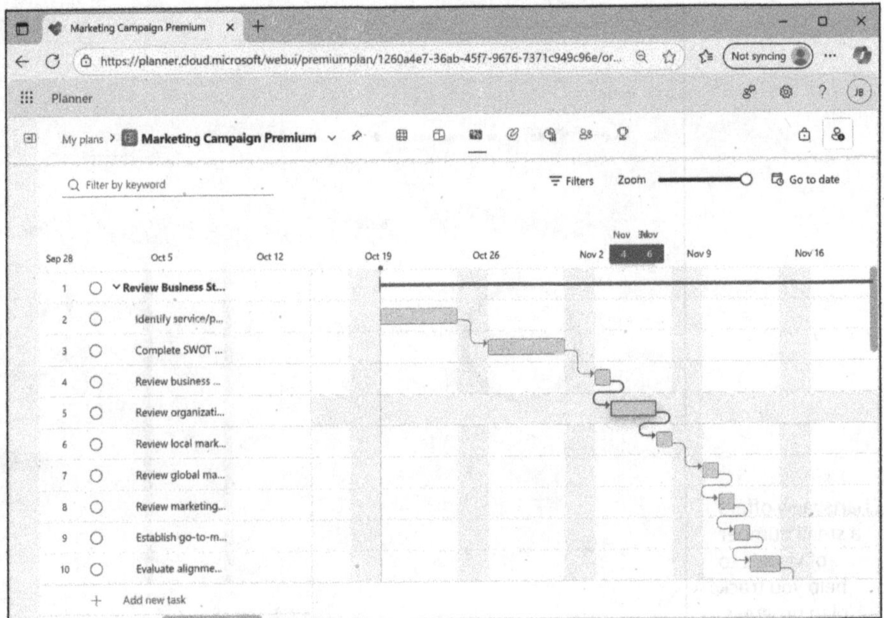

FIGURE 5-16:
Use Timeline view
in Premium plans
to visualize task
dependencies
and timelines.

Many projects will span months or even years, and often you'll need to see as much of the plan as possible. Although you could change the zoom factor of your browser to show a very long span of time, you wouldn't be able to read the information in the view very well (or at all) because the whole interface changes. Instead, you can use the Zoom slider at the upper right of the view to change the span of time shown in the timeline without affecting task names, dates, or other interface elements. You can also use the Go to Date control to jump to a specific date in the plan.

Taking Advantage of Teams Integration and the Mobile App

Other sections of this chapter describe Planner's web interface, which you access through `https://planner.cloud.microsoft`. The web interface currently gives you access to most Planner features. However, a small number of features are currently available only through Teams. In addition, the Planner mobile app offers the means to interact with plans on your mobile device. The following sections explore these two topics.

Planner supports most current web browsers across a range of operating systems including Windows, macOS, Linux, iOS, and Android. Older browser versions may not be supported, or features could be limited. (See Chapter 4 for details on browser compatibility.)

Planner in Teams

You can add Planner as a tab in Teams, which is handy if you and your team use Teams for chat, document sharing, and other collaboration capabilities. The interface is much the same, and using Planner in either Teams or the web app is seamless, but there are a few considerations when choosing between the two:

>> **Additional features:** The Teams interface for Planner currently supports features not available in the web app. These include real-time notifications from the Project Manager agent when tasks are marked complete and are ready for review, the capability to integrate with custom Teams apps, and access to Premium features not yet available through the web app.

>> **Performance:** The web app will generally load your plans and tasks faster than Teams, but the difference will usually be minor.

>> **Usability:** Using Planner in Teams offers access to the Teams-centric features (a pro), but you can't open Planner in a separate Teams window (a con). When you need to work in Teams and Planner at the same time, open Planner in the web app.

New Planner features will very likely show up first in Teams, followed by the web app and mobile app, because Microsoft is prioritizing Teams integration for Planner.

The Planner mobile app

Microsoft offers an app for your mobile device that enables you to work with plans and tasks when you're away from your computer. The app is supported on Android phones and tablets and on Apple iPhone and iPad. The features are fairly limited, but you can create plans, add tasks, view task status, use grouping to organize task views, and check status in Charts view. The mobile app offers a great way to quickly manipulate plans and tasks when you're on the go. For example, you're discussing a project during a lunch meeting and realize you need to add a task. You can open the mobile app and add the task in case you forget to do so when you get back to the office. Figure 5-17 shows the Planner mobile app.

Search Google Play or the App Store for the Microsoft Planner mobile app if it isn't already installed on your device.

FIGURE 5-17:
Use the Planner mobile app to access plans and tasks from your phone.

Chapter **6**

Creating and Managing Tasks

n Chapter 4, I take you on a tour of the basics of opening Microsoft Planner, creating a new plan, naming the plan, and some other basics. Chapter 5 explores the Planner interface. Now it's time to dig in a little deeper and start adding tasks. Creating and managing tasks is really the key function of Planner, so this chapter lays the foundation for your success.

Adding Tasks and Setting Due Dates

When you condense the features offered by Planner down to a single concept, it becomes clear that it's all about managing tasks. You can think of tasks in different ways. Maybe you're defining the actions that team members will perform for a specific project. Maybe you're creating a chore chart for your kids. Duties, chores, errands, appointments — all these items are ultimately discrete pieces of work that you or someone else needs to complete as part of a plan.

Regardless of what the task is about, each task has certain properties associated with it. For example, a task has a name that identifies it and an assigned priority that specifies the importance of the task in that stream of work. Each task likely has a priority. Some tasks repeat throughout your project. This chapter covers all these topics, but let's get started by adding a task!

Creating the plan

You need a plan to work from before you can create tasks. If you've already created a plan, open it. If you haven't created a plan, follow these steps to start a new basic plan called "Learn About Tasks":

1. **Go to** `https://planner.cloud.microsoft`.

2. **Click My Plans.**

3. **Click the New Plan button near the upper-right corner of the page.**

   The Create New page (shown in Figure 6-1) appears.

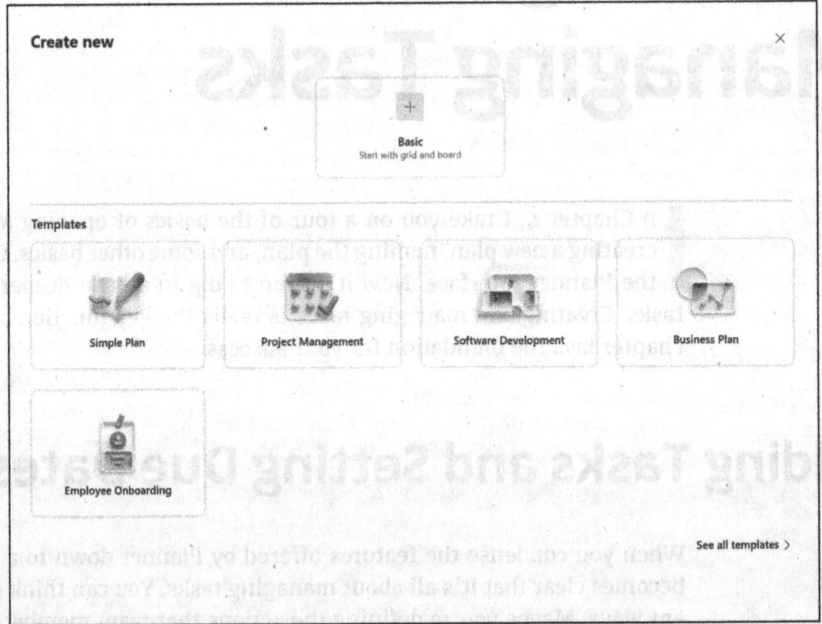

FIGURE 6-1: Choose the Simple Plan template for this example.

4. **Click the Basic button.**

   The Create a Basic Plan from Scratch dialog box appears.

5. **In the Type a Name field, type a name for your plan.**

   I'll use the name Learn About Tasks for this example.

6. **Select the Add to My Pinned Plans check box, and click Create.**

As I explain in Chapter 4, there are a handful of ways to open and interact with Planner. I use `https://planner.cloud.microsoft` to open and use Planner in my examples and let you know when (and why) I'm using a different method.

Figure 6-2 shows the newly created plan, along with any other plans that you've created. Now you're ready to start adding tasks to your plan.

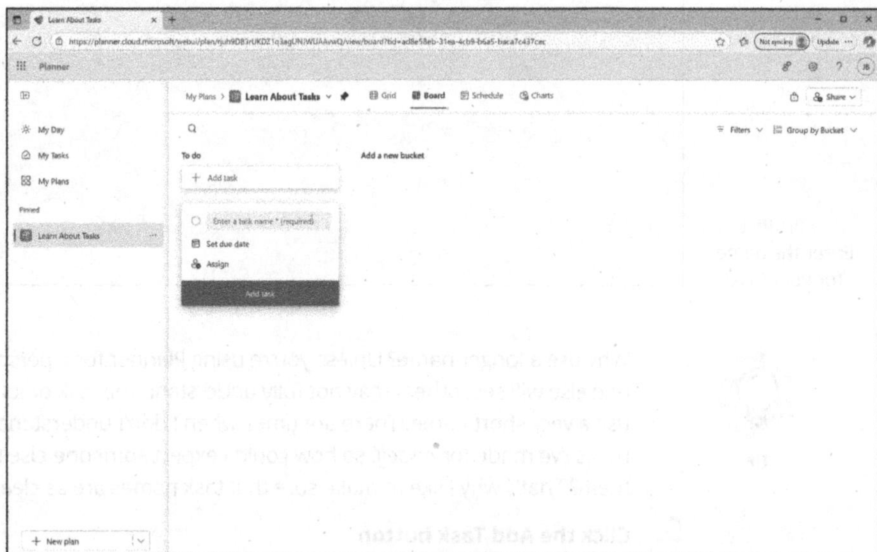

FIGURE 6-2:
Your new simple plan is added to Planner.

Adding a task

Now that you have a basic plan created, you're ready to start adding tasks. That's a simple task (pun intended!). Follow these steps to create your first task:

1. **In the left navigation pane, click My Plans.**

If you pinned the plan (see the preceding section), you can simply click the plan in the Pinned section of the navigation pane.

2. **Click the Learn About Tasks plan that you create in the preceding section.**

3. **If the form for creating the task isn't already showing in Planner, click the Add Task button to display it.**

4. **In the Enter a Task Name box, type** Get Started **(see Figure 6-3).**

When you create a task for real, use a name that identifies as much as possible what the task entails. Instead of, Monthly Report, for example, use Monthly Equipment Status Report Preparation. The name can be up to 255 characters long, so you have plenty of space to be descriptive.

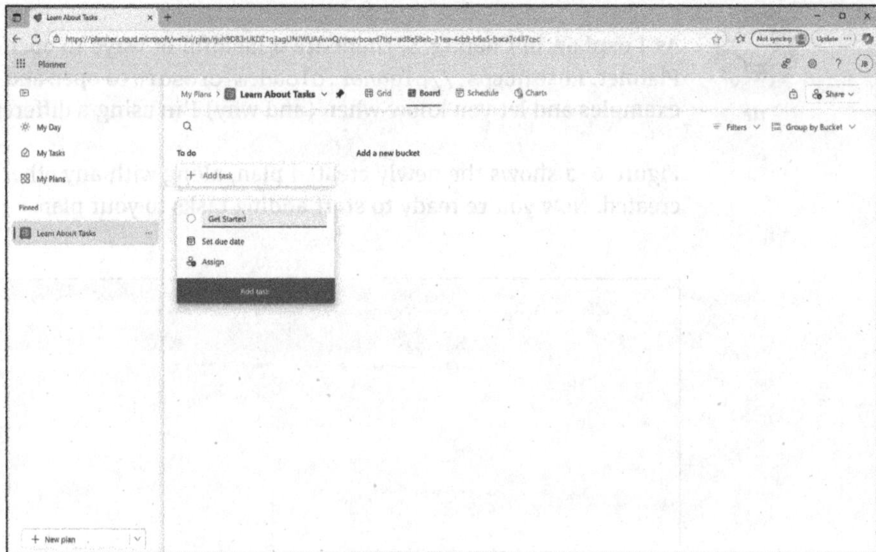

TIP

Why use a longer name? Unless you're using Planner for a personal plan that no one else will see, others may not fully understand the task or its purpose if you use a very short name. There are times when I don't understand some of the notes I've made for *myself,* so how could I expect someone else to understand them? That's why I like to make sure that task names are as clear as possible.

5. Click the Add Task button.

Simple, right? Although you could've added a due date when you created the task, I'll hold that action for the next section. For now, Figure 6-4 shows your new task added to your plan.

At this point, you could begin creating the other tasks in your plan — just follow the same steps. However, you may want to add more context to your tasks, such as setting start dates and due dates, priority, and so on. Read on to learn more about these topics.

Setting the start and due dates

In the early stages of planning, you often won't know when a given task should start or end. Instead, you're thinking about different sections of the project, what top-level tasks will be needed in each section, and how those tasks relate to one another or are sequenced within the project. Unless you've created a draft of the project on the back of a napkin or in some other program (which defeats much of the usefulness of Planner), you're going to flesh out tasks as you move through the initial phases of plan creation. What's more, those tasks will change in various ways as you move through the project.

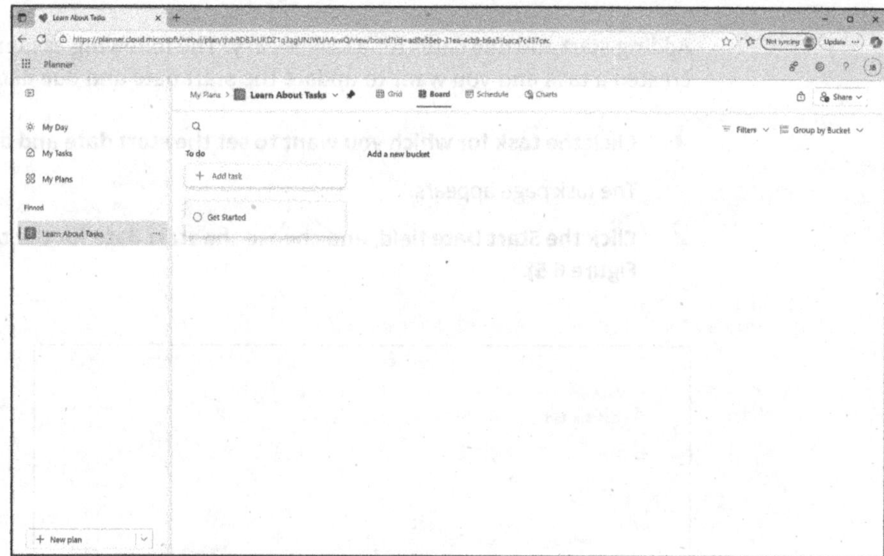

FIGURE 6-4:
The Get Started tasks now appears in the plan.

TIP

I like to start fleshing out a plan by adding top-level tasks (the big rocks that need to be moved), which gives me a sense of the overall scope. Then I begin to organize them into various buckets and add finer detail to the plan. You'll identify additional top-level tasks as you go along and also begin to break up those larger tasks into smaller ones. So, unless you're just adding a task to an existing plan that already has an established timeline, much of the time you'll create tasks without start dates and due dates.

A task can have a start date and a due date. How you use tasks in Planner (or even in a given plan) defines the significance of the start and due dates. For a task assigned to someone to perform one or more actions, the *start date* is the date on which they should start performing those actions. Likewise, the *due date* is when all the actions within the task should be completed to stay on your defined schedule. But you may also use tasks as markers for key activities that aren't actionable by anyone on your team. For example, maybe you need to receive product installation documentation from a manufacturer for equipment being installed in a building project. The due date is when you need to receive the documentation, but the start date could either be the date you place the order for the equipment, the date you notify the manufacturer of which documentation is required, or something else. Regardless, the due date defines your requirement to have the documentation in hand, and the start date simply defines the span of time need for that process to start and complete.

Adding start and due dates for a task is easy. The following assumes you've already created a task and you want to update the start date and due date:

1. **Click the task for which you want to set the start date and due date.**

   The task page appears.

2. **Click the Start Date field, and choose the start date for the task (see Figure 6-5).**

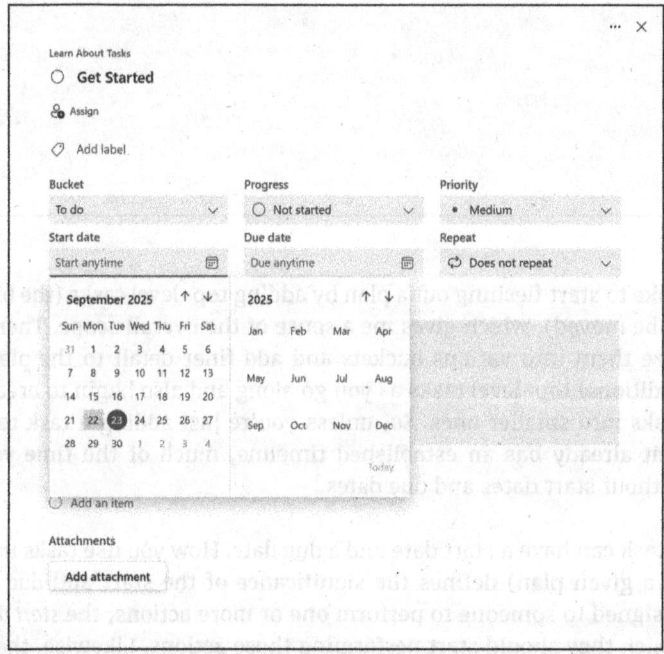

FIGURE 6-5:
Choose a date
using the
date picker.

3. **Click the Due Date field, and choose the date by which the task needs to completed.**

4. **Close the task page by clicking the X at the top right of the page, or simply click anywhere outside the page.**

There isn't an Undo command in Planner — you can't simply press Ctrl+Z to restore the previous date after you've changed it. Instead, you must choose the desired date again.

Although there isn't an Undo command in Planner, you can undo certain changes. For example, if you accidentally delete the notes from a task, you *can* press Ctrl+Z to bring them back. Experiment a bit to get comfortable with what you can and can't undo in the Planner interface.

Setting Priority

Not all tasks are created equal; some are more important than others in the grand scheme of your plan. For example, within a body of work in your plan, you may have a dozen or so tasks that need to be completed to move the project forward, but they may not be *showstoppers* — tasks that absolutely must be completed before others can begin or be completed themselves. For example, putting the finishing touches on the design of a product announcement is less important (has lower priority) than finishing the product itself.

Priority in the context of Planner gives you a means of identifying the importance of a task within the plan. In the free version of Planner and in Planner Plan 1, the options you have for task priority are Low, Medium, High, and Urgent. How you use these priorities within a plan depends on the plan itself and other considerations — in other words, the priorities are *situational.* I can't tell you what tasks should be set to a given priority because Planner can be used in so many ways and scenarios. For example, only you can determine what is a critical priority with a particular plan in your situation. So, as you begin to structure your plan, give some thought to what priority levels should apply to the tasks in the plan.

It's a good idea to define your priorities officially when you create a plan, particularly when others are part of the project and using the plan. Document what each priority signifies, with examples where possible, so that each member of the team has a clear understanding of what's expected.

Task priority and *dependencies* go hand in hand. Task dependency enables you to logically link tasks that depend on one another. For example, the installer for a piece of equipment in your building project probably can't complete the installation of the equipment until they receive the installation specifications from the manufacturer. Even if they've had experience with that equipment in the past, the installer should still have the specifications in hand to ensure that the equipment is installed to the *current* specifications. In this scenario, the completion of the installation task is *dependent* on the completion of the task that results in the documentation being delivered. (I cover dependencies in Chapter 10.)

Now that you have some background in priorities and how you may apply them to tasks, you're ready to set some priorities! Follow these steps:

1. **Open the plan containing the tasks whose priorities you want to set.**

2. **Click a task.**

3. **From the Priority drop-down menu, choose the priority level you want to assign to the task.**

4. **Click the X to close the task or click outside of the task page.**

The task should now display a priority icon in Grid view and Board view (see Figure 6-6).

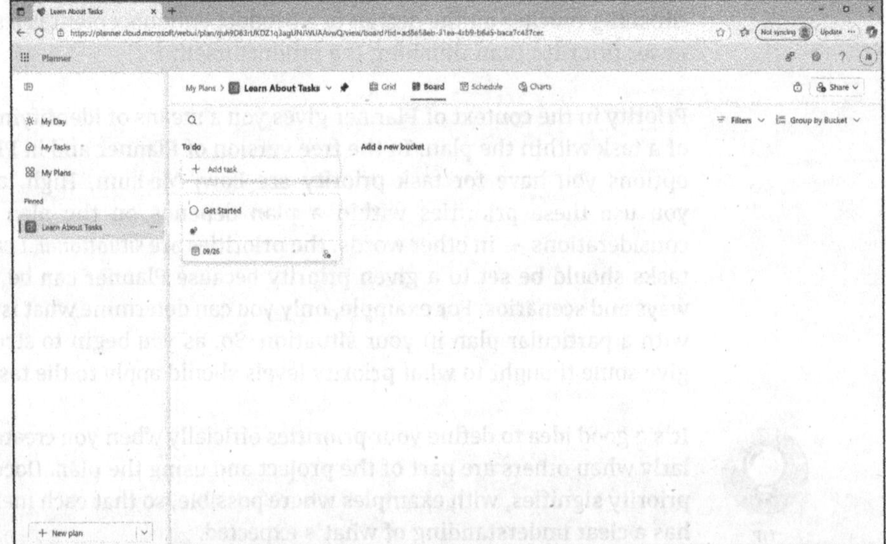

FIGURE 6-6:
Task priorities are represented by icons in the Board view.

Planner assigns Medium priority to any tasks for which you don't specify a priority. You can change that priority any time. Just open the task, change the priority, and close the task. The icon will change accordingly.

Repeating Tasks

When you build a plan, you'll find that sometimes a task needs to be completed multiple times throughout a project. For example, you may have a sign-off process for each phase of a project where one or more people need to agree that that phase is complete before you move to the next phase. Or, maybe you have a weekly team check-in meeting, a monthly report that must be submitted, a recurring maintenance item, or other tasks that need to happen on a repeating basis. Simply put, these are *repeating tasks*.

You can use any of five predefined recurrences in Planner:

>> **Daily:** The task repeats every day.

>> **Weekdays:** The task repeats every weekday (Monday through Friday), but not on weekends.

>> **Weekly:** The task repeats on the same day each week based on the day of the week of the task's start date. Create the initial task for Monday, for example, and the task will repeat every Monday.

>> **Monthly:** The task repeats monthly on the same date as the first instance. For example, if you create the first instance for October 21, the task will recur on the twenty-first of each month after that.

>> **Yearly:** The task repeats annually on the same date as the first instance. For example, if you create a task for January 2 of this year, the next occurrence will be on January 2 of next year.

If you spend a lot of your time in Microsoft Outlook and are familiar with recurring calendar appointments, you're probably wondering, "How do I get more control over these recurrences? I need to repeat a task on the third Wednesday of every month." That's easy — just use the Custom repeat option. This option lets you control the recurrences on a more granular level.

TIP

If you create a task that repeats on the thirtieth of each month, you may wonder what happens in February. I wondered the same thing! What happens is that Planner is smart enough to set the date for the February task to the last day of February instead of the thirtieth. If you truly want a task to repeat on the last day of each month, regardless of the number of days in the month, use the Custom repeat option, explained in more detail later in this section.

Let's work through an example to illustrate the predefined options. Create a one-day task that repeats every week on the second of the month:

1. **Open the plan that will contain the repeating task.**

2. **Create a new task called Monthly Kick-Off Meeting and set a due date for the second of the current month.**

3. **Open the Details page for the task by clicking the task name in the Board view, click the Repeat drop-down list, and choose Monthly (see Figure 6-7).**

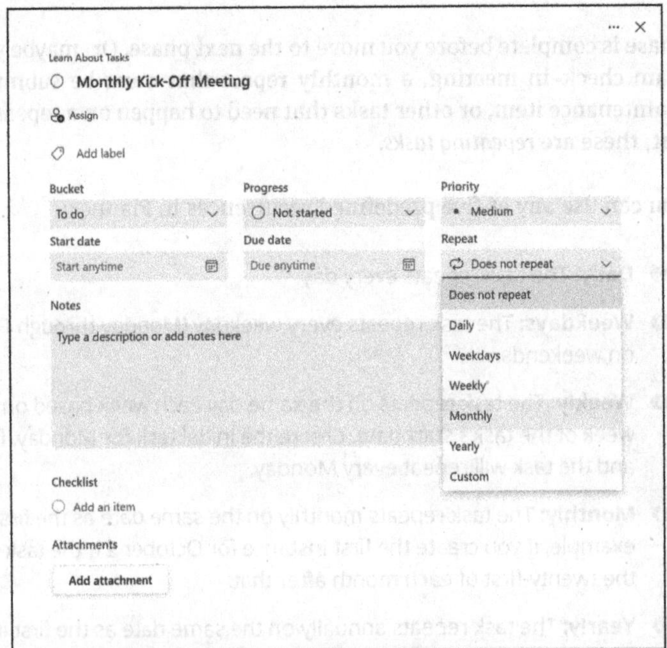

FIGURE 6-7:
Use the Repeat drop-down list to choose the repeat option.

4. **Close the task.**

5. **Click Schedule to open Schedule view.**

6. **Click through future months to see that the task repeats on the second of each month, regardless of the day of the week (see Figure 6-8).**

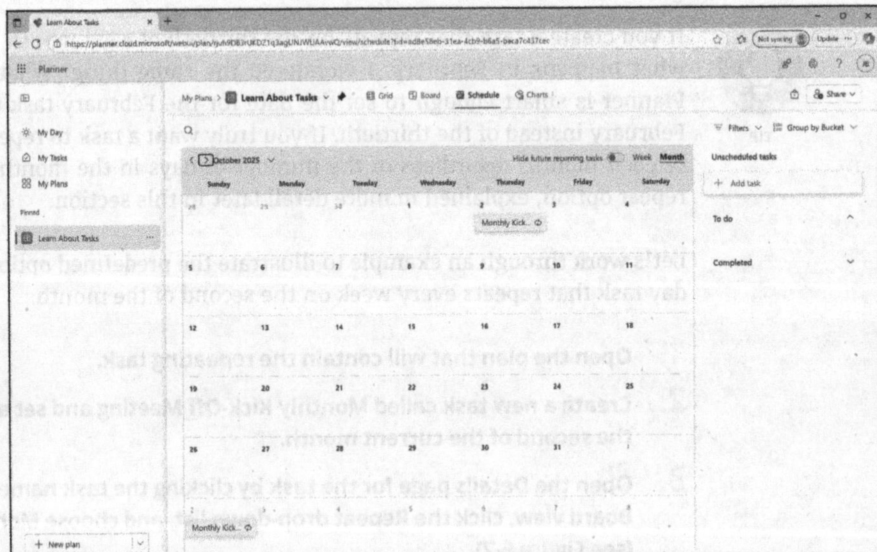

FIGURE 6-8:
Scroll through Schedule view to see your repeating task.

How you open a task's Details page depends on the view you're using. In Board view, you can simply click the task name to open the Details page. In Grid and Timeline views, click More Options (the vertical dots to the right of the task name) and choose Open Details.

The other predefined repeat options work much the same way. Just create the task and set the start date and due date to whatever recurrence you want based on the requirements of the plan and the task itself. Then open the Details page for the task and choose one of the predefined repeat options.

To this point, I've only focused on the due date, and the previous example created a one-day task on the second of each month. However, many tasks will span multiple days or even weeks. A key concept to understand here is that Planner repeats the entire task *time span*, not just the due date. For example, assume that you have a task that needs to span the first through the third of each month. In that situation, you would create a task with a start date on the first day of the month and the due date on the third day of the month. Then set the task to repeat monthly. The result is a task that occurs from the first through the third of each month.

Now let's look at the Custom option, adding a task that repeats on the first Monday of each month:

1. **Create a new task named Custom Repeat.**

2. **Open the Details page for the task and click the Repeat drop-down list (see Figure 6-9).**

3. **Click Custom.**

   The Custom Repeat dialog box (shown in Figure 6-10) appears.

4. **In the Repeat Every text box, enter** 1.

5. **Choose Week from the drop-down list.**

6. **Click M to choose Monday, and if other days are highlighted, click them to deselect those days.**

7. **Click Save.**

I hope these examples have given you enough background to understand the options you have for creating recurring tasks. Spend some time creating tasks with various repeat options and scroll through Schedule view to see the effect of each option.

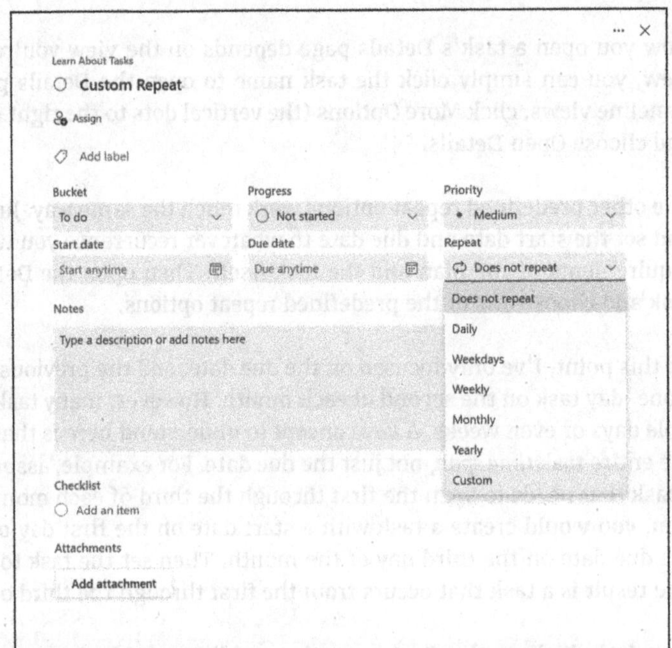

FIGURE 6-9:
Use the Custom
option to gain
more control over
how tasks repeat.

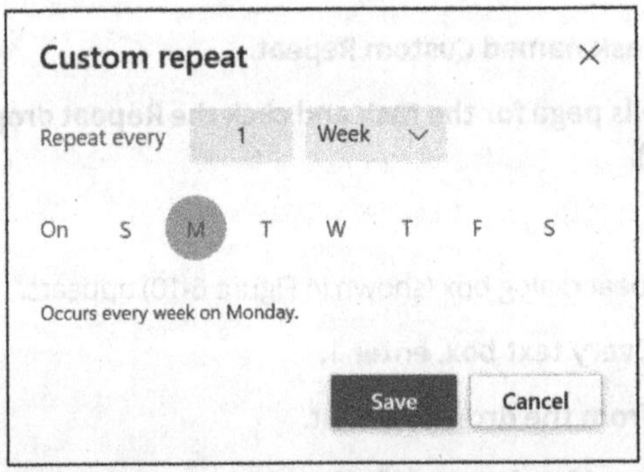

FIGURE 6-10:
Specify your
repeat options
in the Custom
Repeat
dialog box.

Adding Notes

To ensure that you and others are able to not only understand the purpose and desired result of a task but also add information as you work through a task and plan, use Planner's Notes feature.

Entering notes

Figure 6-11 shows the Details page for a task, where you'll find the Notes field. You can enter up to 1,000 characters in the Notes field, enabling you to add a summary of the task activities, add an explanation of expectations, or keep track of task status by adding notes to the task.

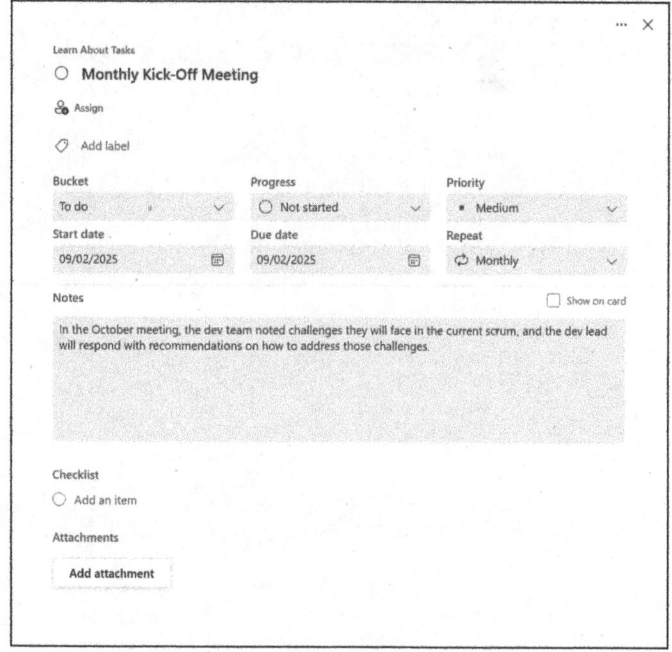

FIGURE 6-11: Use the Notes field to add explanations, expectations, and status information to the task.

Entering notes is simple so I won't run through an example. Just type your notes and close the Details page.

Showing notes on cards

This feature is minor, but it's worth mentioning. On the Details page, note the Show On Card check box above the Notes field. By default, you'll see the note on the Details page but not in Board view. Place a check in the box if you want the notes to show up in Card view (see Figure 6-12).

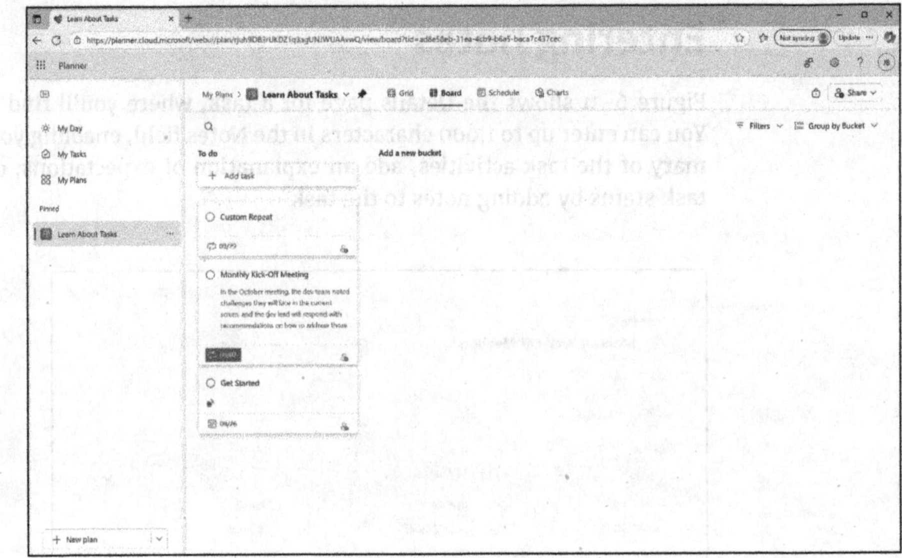

FIGURE 6-12:
You can have the notes for a task show up on the task's card in Card view.

Chapter **7**

Organizing and Structuring Tasks

P roviding a means to organize and structure a project is a key capability of any good planning tool. Without that capability, your project is just a collection of disparate tasks with no interrelationships or logic. That's a disaster in the making, not a project.

Structuring and organizing a project requires more than just putting tasks into buckets and placing them on a timeline. Those are certainly key aspects, but the end goal is not only to bring logical order to the project, but also to make it easy to visualize, understand, and execute. I cover these topics throughout this chapter, but I start with the foundation: organizing your tasks.

Working with Buckets

In Microsoft Planner, *buckets* provide a means of organizing the project's tasks. Personally, I'm not sure that the bucket is a great analogy, but it does offer a visual metaphor for grouping tasks into distinct groups to make it easier to sort, prioritize, and manage tasks. So, although Planner calls them buckets, you can internalize the concept in whatever way makes the most sense to you.

Here are some key reasons and ways to use buckets in Planner:

>> **Visualizing related tasks:** Use buckets to group together related tasks, providing a visual way to organize tasks and bring structure to a project. Tasks in a bucket don't require similar timelines but instead could include tasks from the beginning and the end of the project. For example, all your documentation tasks for a project could fall into one bucket, regardless of when they're due. The same is true for all communication tasks and other logically related tasks. The key is to use buckets to provide a *logical* grouping of tasks.

>> **Categorizing:** Another way to think about task organization is by task category. This can include tasks that fall into the same phase of a project, tasks assigned to a particular team, tasks with a particular priority, and so on. Don't just consider the outcome of a task to decide what bucket it should be in; also think about who will accomplish the task, how important it is, when it occurs in the project, and so on.

>> **Focusing and prioritizing:** Using buckets to organize tasks in Planner helps you and the team focus on specific groups of tasks. Without buckets, all you have is a big to-do list with little logical order. Grouping them in a logical way that fits the project and the way you or your team works helps you focus and prioritize tasks.

>> **Tracking:** Just as buckets help you focus and drive specific groups of tasks, they also help you track progress and completion of those tasks. Using the previous documentation example, the team responsible for that set of tasks can use checklists and the status of those checklists to quickly see completion status of all the project's documentation.

>> **Collaborating:** Buckets offer a great means of aligning a set of tasks to a team and enabling the team to assign and complete their tasks in one place. Using buckets in this way clarifies responsibilities and streamlines collaboration with a team, as well as across teams.

Visually, buckets appear in Board view when the view is grouped by bucket, as shown in Figure 7-1. The buckets are displayed as a set of columns with the tasks listed underneath. In this example, the visible buckets are named Initiating, Planning, and Executing.

Grouping by bucket is just one option for displaying tasks in Board view. To see the other options, click the Group by Bucket drop-down list and choose the property you want to use to group the tasks. For example, select Priority to view the tasks grouped into Urgent, Important, Medium, and Low columns.

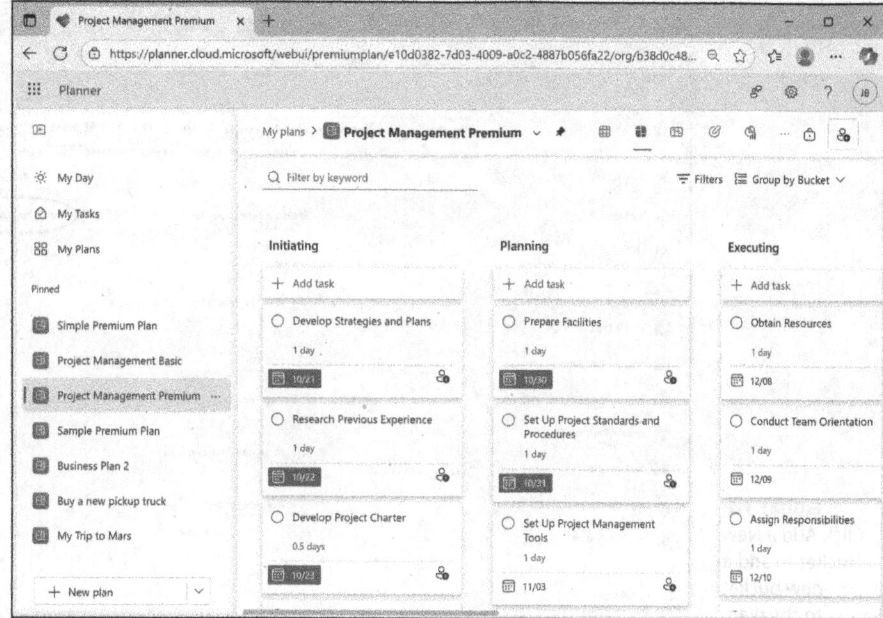

FIGURE 7-1: Buckets enable you to organize tasks in Planner.

Creating and removing buckets

Creating new buckets is easy. Follow these steps to create one:

1. **Open the plan in which you want to create the bucket.**

2. **Click Board on the toolbar to open the Board view.**

3. **Scroll to the right and click Add a New Bucket (see Figure 7-2).**

4. **In the Bucket Name field, enter the name of the bucket.**

5. **Click outside the text box or press Enter.**

As you work to organize a plan, occasionally you'll want to remove buckets. Maybe the way you think about the plan has changed and you want a different organizational structure. Or maybe you just created a few extra buckets during plan setup and decided you don't need them. In any case, it's easy to remove a bucket. Follow these steps:

1. **Open the plan containing the bucket you want to remove.**

2. **Click Board in the toolbar to open the Board view.**

3. **Click the Group By button and choose Group by Bucket.**

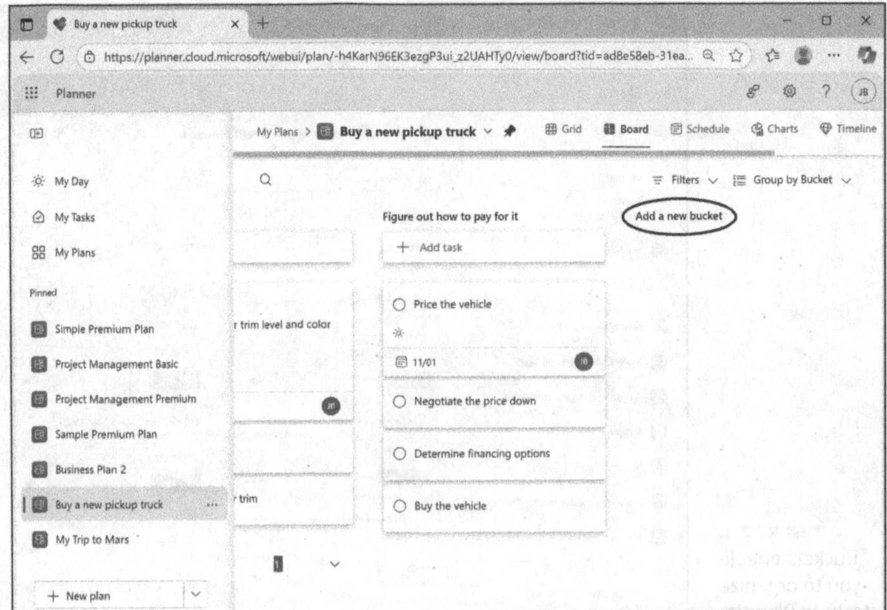

FIGURE 7-2:
Click Add a New Bucket to add a new bucket to the plan.

4. **Hover the mouse over the bucket name and click the ellipsis that appears at the right corner of the bucket header.**

5. **Choose Delete.**

   If there are tasks in the bucket, Planner warns you that deleting the bucket will also delete all the tasks. Click Delete if you want to delete the tasks. Otherwise, click Cancel, move the tasks to a different bucket, and then delete the bucket.

WARNING

After you delete a task and/or a bucket, those items are generally not recoverable. The exception is if you've implemented a backup and recovery solution for Planner. (See Chapter 4 to explore backup and restore options for Planner.)

Coloring buckets

PREMIUM

You have a bit of flexibility in how you set up the buckets in a plan. Naming the bucket is one example, but you can also change the color Planner uses to display the bucket if you're editing a Premium plan. For example, maybe you want to highlight a group of critical tasks. One way to do that is to set the bucket color to red. Similarly, you may want to color green the buckets where all the tasks are complete.

Changing color is easy. Follow these steps:

1. **Open the plan in which you want to modify the bucket.**

2. **Hover the mouse over the bucket's name.**

3. **Click the Color Picker icon, and then choose a color from the resulting list.**

Figure 7-3 shows a bucket that I've colored.

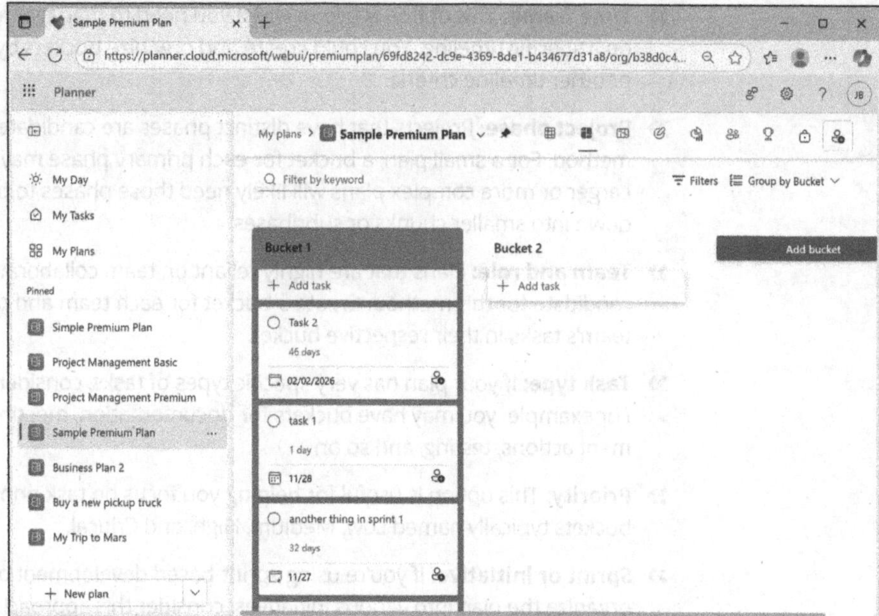

FIGURE 7-3:
Add color to
buckets to
highlight tasks.

Structuring work with buckets

The whole purpose of buckets is to give you a way of organizing tasks that makes sense for each plan. So, the buckets in one plan could be completely different from the buckets in another plan. Your first step is to consider the tasks the plan will encompass and decide how you want to group them within the logical construct of the plan. (I offer some suggestions for ways to use buckets for task organization in the "Working with Buckets" section, earlier in this chapter.)

The following recommendations apply primarily to plans for which you create the structure yourself instead of using a template. Even so, you can use the suggestions to fine-tune a template-based plan's structure.

There is no right or wrong way to organize buckets in Planner, but there are several conceptual methods to consider:

>> **Workflow stage:** Use (and name) buckets to represent the plan's workflow. For example, you may create buckets named Not Started, In Progress, In Review, and Completed. This method implies that you'll move tasks between buckets as their status changes, either manually or with Microsoft Power Automate. I suggest using this method only for smaller, simpler plans because of the need to move tasks.

>> **Time frame:** This option is useful when you need to manage the plan based on a specific timeline. You could create and organize buckets by week, month, or other timeline criteria.

>> **Project phase:** Projects that have distinct phases are candidates for this method. For a small plan, a bucket for each primary phase may make sense. Larger or more complex plans will likely need those phases to be broken down into smaller chunks or subphases.

>> **Team and role:** Plans that are highly reliant on team collaboration are a good candidate for this method. Create a bucket for each team and place each team's tasks in their respective bucket.

>> **Task type:** If your plan has very specific types of tasks, consider this option. For example, you may have buckets for documentation, meetings, development actions, testing, and so on.

>> **Priority:** This option is useful for helping you focus on task priority with buckets typically named Low, Medium, High, and Critical.

>> **Sprint or initiative:** If you're using sprint-based development or you want to organize the plan into various initiatives, consider this approach.

As you read through those descriptions, you may have thought that some are too restrictive and wouldn't provide a good way to organize or visualize your task timeline. Keep in mind that even though you can use buckets to organize tasks by timeline or phase, the main reason to use buckets is to *organize* tasks in a logical way, not necessarily visualize them across the plan's timeline. You can use the Schedule and Timeline (Premium) views for visualizing them.

After you've decided on a working structure, begin creating the buckets that fit that structure using these points as a guide:

>> Name buckets in such a way that everyone who works with the plan can understand the intent of the buckets and the types of tasks they contain.

» Create the buckets before you start to add tasks. This helps you focus on the plan's structure before you go deeper into individual tasks.

» Use color to call out specific buckets and tasks based on team assignment, priority, or other criteria as suits the plan and the way you work. Be consistent and make sure the purpose of the colors you assign will be easily recognized by your team. Keep it simple and use colors sparingly.

» Organize the placement of buckets in Board view, moving them left or right to visually highlight a timeline or other logical flow. To move a bucket left or right, open Board view, hover the mouse pointer over the bucket's header, click the ellipsis in the bucket's header, and click either Move Right or Move Left.

Categorizing Tasks with Labels

Whether you're working with a Basic or Premium plan, Planner offers the option of adding labels to tasks. Planner includes 25 predefined color-based labels that you can use to quickly visualize tasks. For example, you may use labels to help you quickly identify tasks by type, priority, team or role assignment, project or initiative, sensitivity, and so on. You can also use labels to indicate special criteria for tasks, such as those that need approval, are blocked, or have other special requirements.

Labels do more than just add visual clues for tasks, however. You can also use them to focus on specific tasks through filtering and reporting. For example, let's say you assigned a Needs Approval label to several tasks throughout your plan and you want to view all those tasks to check status. In that scenario, you can filter Grid view to show only tasks that have that label to see which ones still need approval.

To view tasks that have a specific label, click the Filters drop-down menu in the view and choose Labels followed by the desired label.

Assigning and editing labels

Like most actions in Planner, adding a label to a task is easy. Follow these steps:

1. Find the task in Planner, and then click the ellipsis at the right of the task.

2. Click Label and then choose a label from the list (see Figure 7-4).

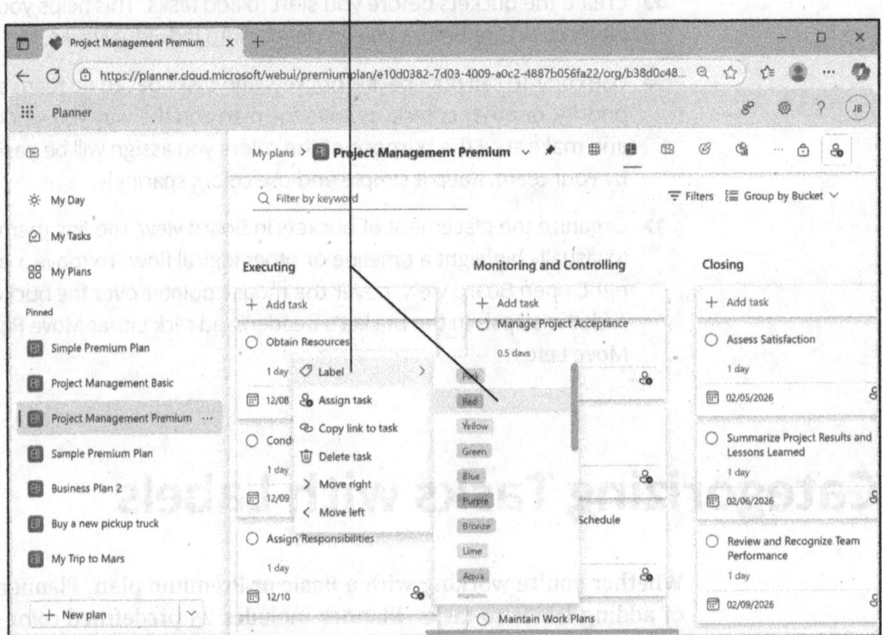

FIGURE 7-4:
Apply labels to tasks for visualization and filtering.

TECHNICAL STUFF

If you're familiar with categories in Microsoft Outlook, you'll have no trouble understanding and using labels because they function much the same. By default, Outlook categories have a color and a name that matches the color. You can edit those categories or create new categories using other names like Reactive, Important, Customer Email, and so on, each with an identifying color. The main difference between categories in Outlook and labels in Planner is that you're limited to 25 labels per plan, whereas Outlook doesn't impose a category limit.

What about situations where you need more than one label? No problem! You can assign as many of the 25 labels to a task as necessary. For example, you may assign a Priority label and a Needs Approval label to a task. Both labels then appear with the task.

TIP

When you filter a view by label, Planner is not exclusive in deciding what tasks to show in the view — it displays the tasks that have the selected label regardless of whether they also have others. For example, if you have a task named Purchase Servers with the labels Needs Approval and High Priority and you filter the view to show those tasks with the label Needs Approval, the Purchase Servers tasks will appear in the view.

You may be a bit puzzled the first time you try to remove a label that you've assigned to a task in Planner. I was! I expected a No Label option in the label list, but there isn't one. Instead, you toggle the label on or off. For example, if you've assigned the Red label to a task (indicated by a check mark next to the label name in the drop-down list), you can remove the label simply by clicking the Red label again.

You can also delete a label from a task's Details page. Just click the X beside the label.

Creating your own labels

The colored labels in Planner are handy, but they aren't very descriptive with just a color for a name, particularly when you want to identify priority tasks, those that need approvals, and so on. Fortunately, you can customize the labels to suit your needs for a given plan.

Keep these factors in mind when customizing labels in Planner:

>> You're limited to 25 labels per plan.

>> If you modify a label name, all tasks that have that label attached will change to reflect the new label name.

>> You can't change the color of a label or add other colors. Planner gives you 25 colors to work with, and all you can change is the label name. So, plan your colors ahead of time whenever possible and note which color will have a specific name. For example, you may want to reserve the Red or Cranberry label to rename for your Priority or Overdue task label.

Follow these steps to create custom labels:

1. **Create a list of the labels you want to use and what color they'll be.**

2. **Open the Details page for the task that will have the custom label.**

   How you open a task's Details page depends on the view you're using. In Board view, you can simply click the task name to open the Details page. In Grid and Timeline views, click More Options (the vertical dots to the right of the task name) and choose Open Details.

3. **Click Add Label (see Figure 7-5).**

4. **Scroll through the list to find the color you want to use, and then click the pencil icon beside that color.**

   The Edit Label text box appears (see Figure 7-6).

5. **Type the name of the label.**

6. **Add or modify other task labels as needed, and then click the X in the top right corner to close Details page.**

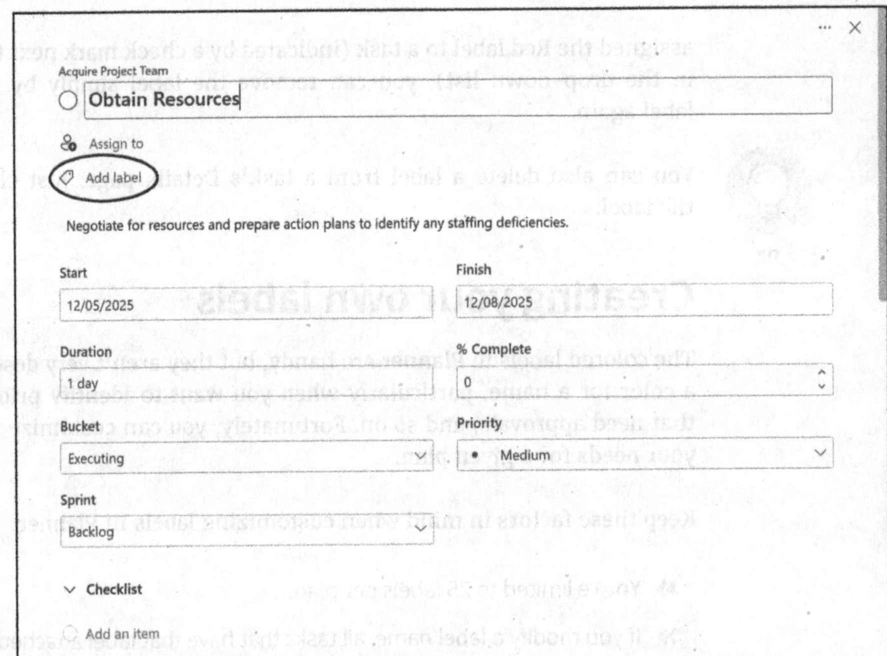

FIGURE 7-5:
Click Add Label
to choose an
existing label.

Type a new name for the label

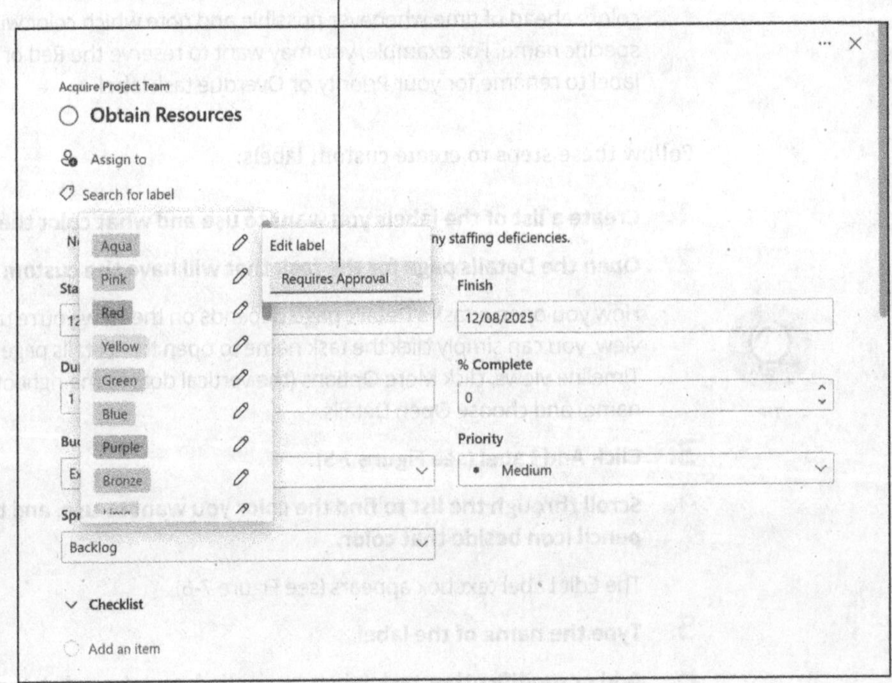

FIGURE 7-6:
Choose a color
and then edit a
label name to
create your
own label.

You don't need to assign the labels when you create them. For example, I often open a new task, create the labels needed for the plan, and then delete the task from the plan. But you can also use an existing task. If you don't want to assign the new labels to the task, just delete them from the Details page before closing it.

Using Checklists

A task isn't necessarily a stand-alone activity. Let's use mowing the lawn as an example. You could create a task for your weekend plan called "Mow the lawn." You probably don't see that as a set of subtasks that you have to perform, but those subtasks can include getting gas for the mower or charging its batteries, picking up twigs or branches that fell during the last big wind, sharpening the blades, moving toys out of the way, and trimming the edges. Mowing the lawn just got more complicated! Fortunately, Planner offers a checklist feature that can help.

Understanding checklists

The checklist feature in Planner gives you a way to create and track activities (subtasks) that drive the completion of the overall task. Now that the lawn is mowed, let's take a more business-like example using the purchase of some servers for a new application-hosting project. I'll name the task, Server Acquisition, and the checklist items may be the following:

>> Determine memory and storage specifications.

>> Confirm input/output requirements.

>> Verify required operating system version.

>> Obtain total cost estimate.

>> Receive purchase authorization.

>> Submit purchase requisition to Purchasing.

>> Determine target arrival date.

>> Confirm complete receipt of all items.

The checklist for a task becomes the means for you to understand what steps or actions need to be taken to complete the task. It can also help you track completion. When you finish an action, just click the item in the checklist to mark that action as completed.

One important advantage to using templates when you create a plan is that many of the tasks will already have their checklists defined, which can save you a huge amount of time in building out the plan's structure. They can also provide a starting point from which to begin modifying the checklist, adding steps that are specific to your team, organization, or planning style.

Creating a checklist

As in most aspects of creating a plan, you should spend some time thinking about the checklist items that you'll add to each task. You may do that on a notepad or right in Planner. You may have documents that you use as a guide to create the checklist. Whatever method you use to get ready, you'll find it's easy to add the checklist when the time comes. The following steps explain how, using the Server Acquisition task I describe in the preceding section:

1. **Create the Server Acquisition task.**

2. **Open the task's Details page.**

3. **Scroll down to find the Checklist section (see Figure 7-7).**

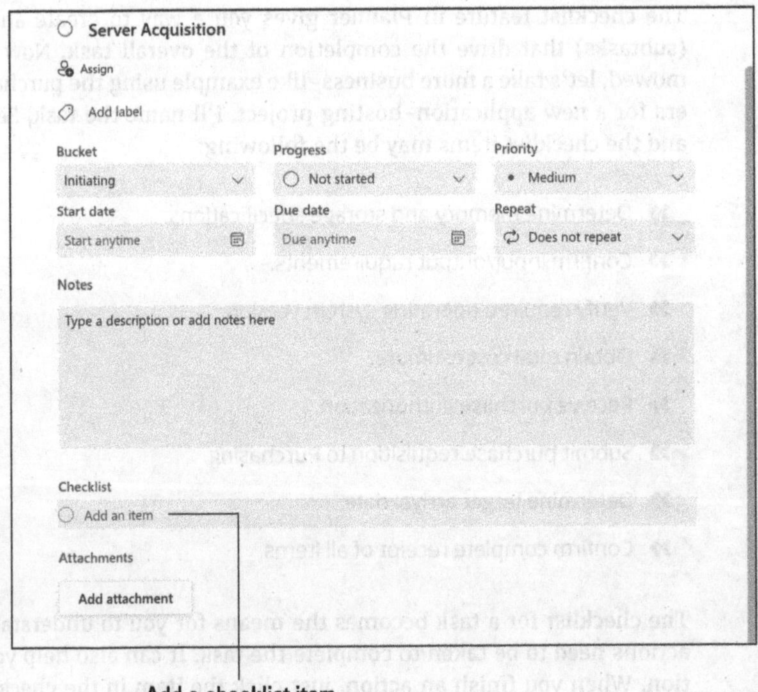

FIGURE 7-7: Create a checklist in the task's Details page.

4. **Click the Add an Item text box, type** Determine memory and storage requirements, **and press Enter.**

5. **Continue adding the other checklist items by typing the checklist name and pressing Enter.**

6. **When the checklist is complete, click the X in the upper-right corner to close the Details page.**

The checklist items appear for all tasks and types of plans on the task's Details page, as shown in Figure 7-7. On Basic plans, you also have the option to show the checklist on the task card in Board view, along with an indicator showing how many of the checklist items are complete. Figure 7-8 shows an example. (Not all checklist items may appear on the task card, depending on the number of items in the list.)

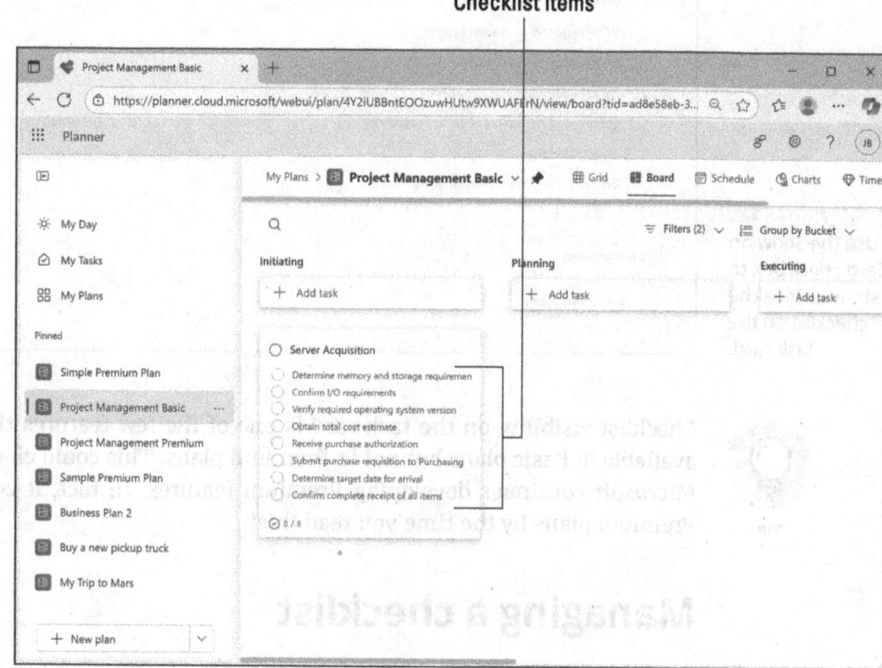

FIGURE 7-8: Basic plans can include the checklist on the task card in Board view.

Follow these steps to add the checklist to the task card:

1. **Open the Details page for the task.**

2. **Scroll down on the Details page to the Checklist section.**

3. **Place a check in the Show on Card check box (see Figure 7-9).**

4. **Click the X in the upper-right corner to close the Details page.**

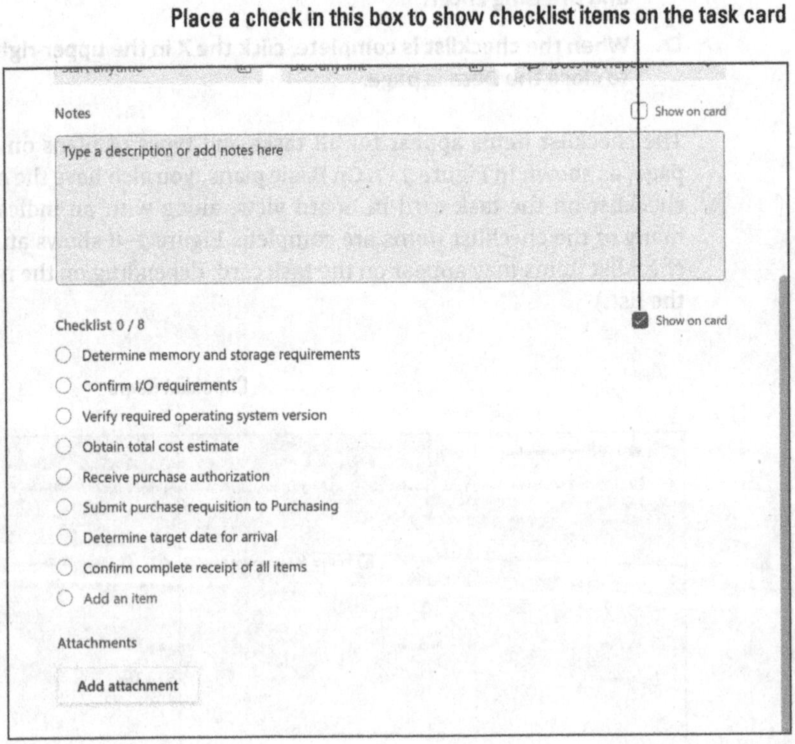

Place a check in this box to show checklist items on the task card

FIGURE 7-9:
Use the Show on
Card check box to
show or hide the
checklist on the
task card.

Checklist visibility on the task card is one of the few features that are currently available in Basic plans but not in Premium plans. This could change over time as Microsoft continues developing Premium features. In fact, it could be added to Premium plans by the time you read this!

Managing a checklist

Working with checklists in Planner is very simple. If you see the checklist on the task card, you can simply click the circle next to an item to mark it as completed. Otherwise, open the Details page for the task and scroll down to the Checklist section. Simply click the circle to the left of an item to mark it as completed.

Chapter **8**

Collaborating and Communicating with Microsoft Planner

Microsoft Planner is a great tool for managing personal or other plans where you're the only participant. Planner also offers a great set of features for collaborating with others on a plan and integrating with other applications, including Microsoft Teams and Microsoft Outlook. Whether you're using Planner just for yourself and you want to access your tasks in Outlook, or you need to collaborate with a team on a project, Planner has you covered. This chapter explores the collaboration and application integration features offered by Planner to support your planning efforts in both scenarios.

Working with Comments

As you work on tasks and an overall plan, you need to be able to track activities and have a good understanding of task status. Comments are one of the tools that you and your team use to provide status updates and add other information to tasks. Notifications can work hand in hand with comments in that Planner can notify you when someone adds comments to a task (see "Setting Notifications," later in this chapter). I save notifications for later and focus in this section on how to add and work with comments in Planner.

Meeting minimum requirements to use comments

There are a few basic requirements for you to use comment functionality in Planner. First, comments require that a plan be part of a Microsoft 365 group, even if you're the only person working with the plan. The Comments field won't show up in the task form if the plan isn't linked to a Microsoft 365 group.

If you have a Planner Basic Plan, comments are stored in the group's conversation history. To enable commenting, you either need to share the plan with a group or assign a task from the plan to someone in the group.

If you have a Planner Premium Plan, Teams conversations are used for comments, so you need to pin the plan to a Teams channel to enable comments.

Planner uses Microsoft Exchange and Outlook group conversations to store comments, which is why a plan needs to be linked to a group. Without that linkage, there is no place to store the comments.

What if you're just using Planner for you own personal projects? You'll still want to add comments to tasks. The answer is simple. Create your own one-person group. This will enable comments in the plan.

But just what is Microsoft 365 Groups? At its core, Microsoft 365 Groups provides an easy means of granting a group of people access to a collection of shared resources, because permissions are automatically assigned to people when you add them to a group. The scope of shared resources can be broad, as indicated in Figure 8-1. In addition to Planner, this includes a shared mailbox, Teams channels, Microsoft SharePoint, Microsoft OneNote, and more.

There are multiple ways to create a Microsoft 365 group. Each method creates the desired group, but there are different results depending on the method you choose. Here's a summary of the methods and implications:

>> **Microsoft 365 Admin Center:** You can create a group through the Admin Center at https://admin.microsoft.com if you have the appropriate permissions in the tenant (the instance of Entra ID that provides identity and authentication services for your organization).

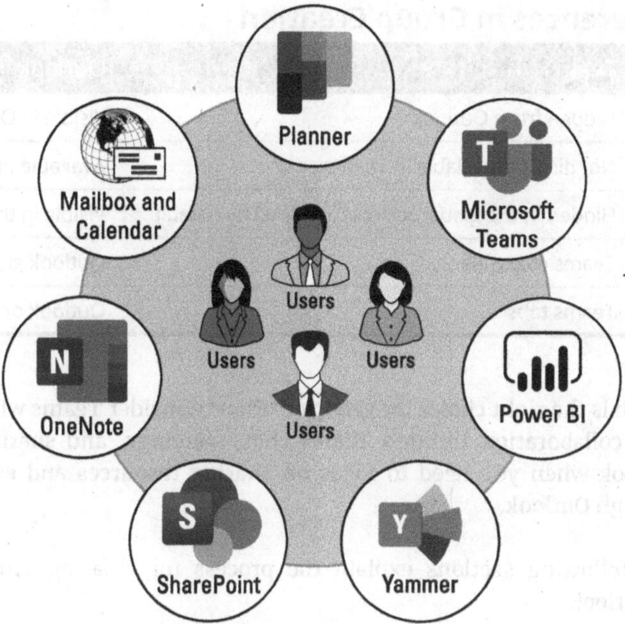

FIGURE 8-1:
A Microsoft 365 group simplifies resource sharing to a group of users.

TECHNICAL STUFF

A *tenant* in Microsoft terms is a secure, dedicated environment for storing and managing user accounts, device accounts, subscriptions, and other data for your organization in Microsoft's cloud. You can think of the tenant as the cloud environment where your user account and your Microsoft 365 apps, including Planner, live in the cloud.

>> **Teams:** Creating a new team automatically creates a Microsoft 365 group associated with the team.

>> **SharePoint:** When you create a site in SharePoint, you can choose to connect it to a Microsoft 365 group.

>> **Outlook:** You can create a group through Outlook Web App at `https://outlook.office.com`.

>> **Exchange Admin:** Exchange administrators can create groups through the Exchange admin portal.

The methods you're most likely to use if you're a user and not an administrator are Teams and Outlook. Keep in mind that there are some differences in results when you choose one of these methods over the other. Table 8-1 explains the differences.

 Differences in Group Creation

Feature	Teams	Outlook
Group mailbox	Hidden from Outlook	Visible in Outlook
Shared calendar	Not directly available in Teams	Available in Outlook
Group email address	Hidden in the global address list (GAL) by default	Visible in the GAL by default
Group conversations	Teams channels	Outlook group mailbox
Planner integration	Teams tabs	Outlook or Planner web app

TIP

Which is the right choice for group creation? Consider Teams when you need real-time collaboration through Teams chat, meetings, and sharing files. Consider Outlook when you need to focus on sharing resources and email collaboration through Outlook.

The following sections explain the process for creating groups using Teams or Outlook.

Creating a group from Teams

Follow these steps to create a group in Teams to use with Planner:

1. **Open Teams and sign in with your user account.**

2. **In the navigation pane, click Teams and Channels.**

3. **In the upper right of the Teams window, click Create Team (see Figure 8-2).**

   The Create a Team dialog box appears (see Figure 8-3).

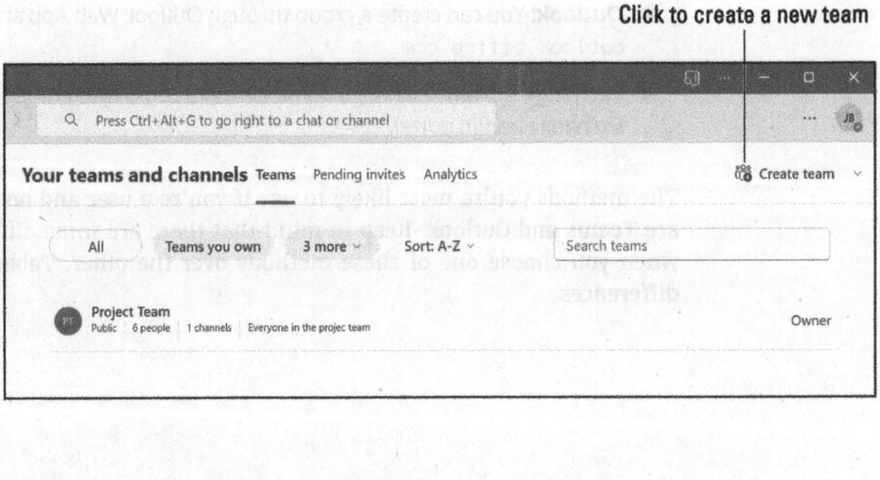

FIGURE 8-2:
Create a
Microsoft 365
group by
creating a team.

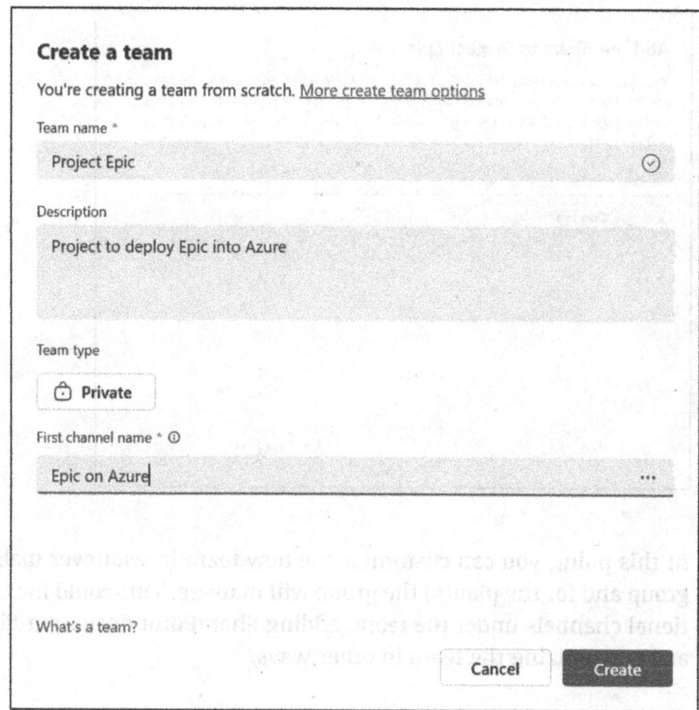

FIGURE 8-3:
Enter the team name and other properties.

4. **In the Team Name field, enter a name.**

5. **In the Description field, enter a description.**

6. **From the Team Type drop-down, select a team type (Private, Public, or Org-Wide).**

   TIP

   In most cases, you should choose Private because your plans are almost always managed by a specific group of people rather than your entire organization. Choosing Private enables you to control who gets into the group and limits the scope of the plan to just the group.

7. **Click Create.**

   Teams prompts you to add members to the team (see Figure 8-4).

8. **Type the name or email of the first person you want to add to the team and select the person from the resulting list.**

9. **Repeat Step 8 until you've added all the initial users to the team and then click Add.**

   The new team appears in Teams.

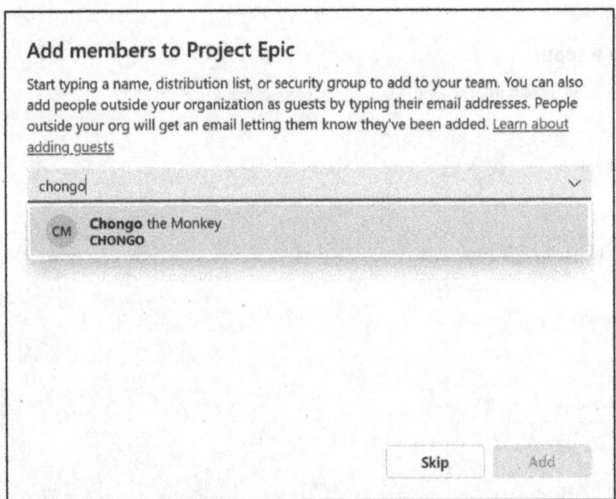

FIGURE 8-4:
Add members
to the team.

At this point, you can customize the new team in whatever makes sense for your
group and for the plan(s) the group will manage. This could include creating additional channels under the team, adding SharePoint document libraries or folders,
and customizing the team in other ways.

Creating a group from Outlook

If your primary focus is collaborating through Outlook and you want to take
advantage of the group's shared mailbox and calendar in Outlook, create the group
from Outlook.

**TECHNICAL
STUFF**

Unless otherwise noted, all Outlook examples in *Microsoft Planner For Dummies* use
Outlook Classic. Why? I've used Outlook for decades, and even written books about
it, and it has a breadth of additional features that the new version of Outlook
doesn't have — at least, for now. Hopefully, the new version of Outlook will eventually have all the features and capabilities of Outlook Classic. Sorry, Microsoft,
I'm not switching until it does.

Here's how to create a group from Outlook:

1. **In Outlook, open the Home tab of the Ribbon.**

2. **Click the arrow beside the New Email control and choose Group
    (see Figure 8-5).**

    The Create Group dialog box appears (see Figure 8-6).

3. **In the Name field, enter the group name.**

4. **In the Email Address field, enter the group email address.**

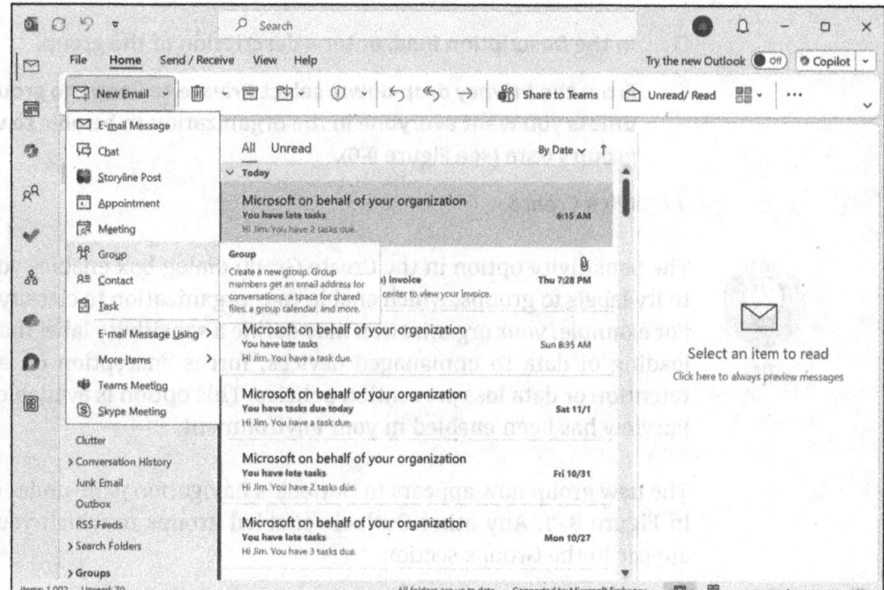

FIGURE 8-5: You can create a Microsoft 365 group from Outlook.

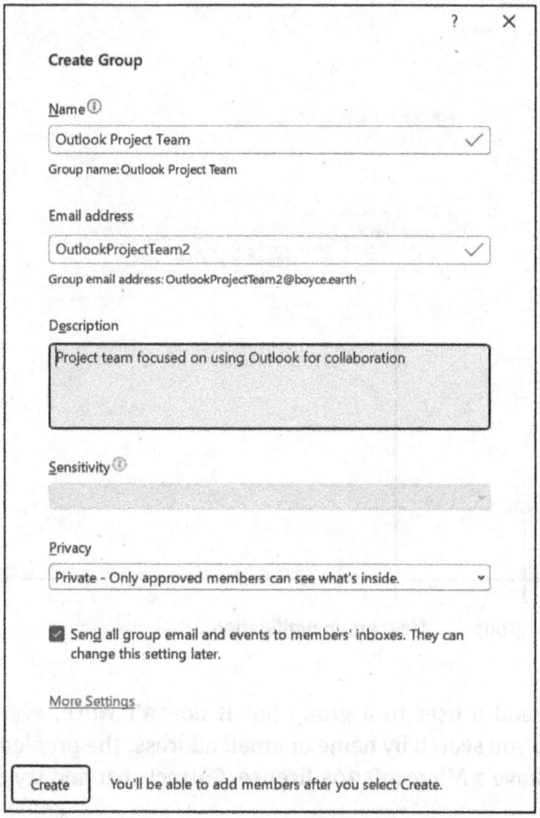

Create Group

Name

Outlook Project Team

Group name: Outlook Project Team

Email address

OutlookProjectTeam2

Group email address: OutlookProjectTeam2@boyce.earth .

Description

Project team focused on using Outlook for collaboration

Sensitivity

Privacy

Private - Only approved members can see what's inside.

☑ Send all group email and events to members' inboxes. They can change this setting later.

More Settings

Create   You'll be able to add members after you select Create.

FIGURE 8-6: Enter the properties for the new group.

5. **In the Description field, enter a description of the group.**

6. **From the Privacy drop-down, select Private to keep the group secure unless you want everyone in the organization to be able to view the group's data (see Figure 8-6).**

7. **Click Create.**

TIP

The Sensitivity option in the Create Group dialog box enables you to apply sensitivity labels to groups, which enable your organization to classify and protect data. For example, your organization may define a sensitivity label that prevents downloading of data to unmanaged devices, forces encryption on emails, or applies retention or data loss prevention policies. This option is available only if Microsoft Purview has been enabled in your environment.

The new group now appears in Outlook's navigation pane under Groups, as shown in Figure 8-7. Any other Outlook-enabled groups to which you belong will also appear in the Groups section.

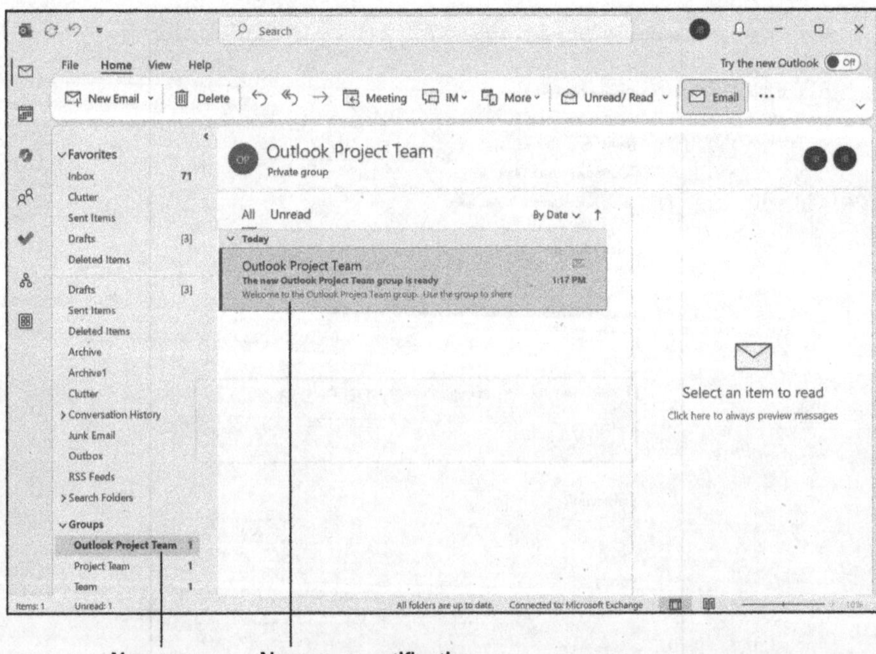

FIGURE 8-7:
Your new group appears under Groups in the Outlook navigation pane.

TIP

If you try to add a user to a group but it doesn't work, even though the user appears when you search by name or email address, the problem could be that the user doesn't have a Microsoft 365 license. Correct that and try again.

Choosing the best of both worlds

What if you want the best of both worlds: full visibility in Outlook but also full functionality in Teams? You can do that, but it takes a slightly different approach that starts in Outlook:

1. **Start by creating the group in Outlook as explained in the preceding section.**

2. **Open Teams, in the navigation pane click the ellipsis beside Teams and Channels, and choose Your Teams and Channels.**

3. **Click Create Team.**

   The Create a Team window appears.

4. **Click More Create Teams Options.**

   The Create Team window appears (see Figure 8-8).

5. **Choose From Group.**

6. **Locate and hover over the group you want to use, and then click Add Group.**

   In this example, I'm connecting Teams to the Outlook Project Team I create in the preceding section.

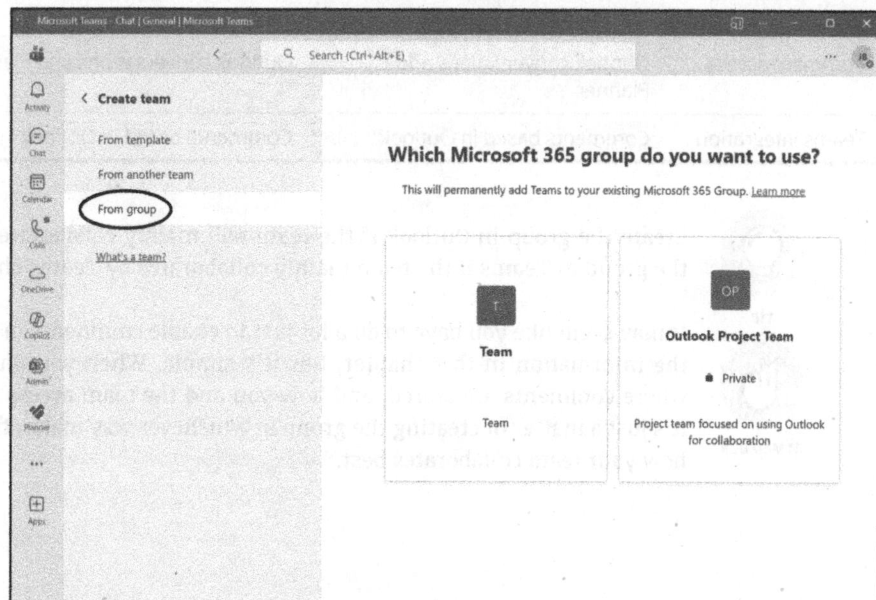

FIGURE 8-8:
The Create
Team window.

You now have a group that gives you the best of both worlds: the collaboration tools in Teams (SharePoint, chat, and so on) and the shared mailbox and calendar in Outlook. Planner will store comments in the group mailbox. Comments are visible through the task details in Planner. Users can view all comments for the plan in the group's conversation thread in Outlook.

"Wait!" you say. "I started in Teams and now I want to add visibility to all that good Outlook stuff." Fear not, this, too, is possible. Your Exchange Online administrator just needs to run a PowerShell script that makes the group mailbox and calendar visible. That script is outside the scope of this book, but if you're an administrator and you need to do that yourself, a quick internet search should give you the information you need.

You may be confused at this point. To be honest, I was confused the first time I worked through these two scenarios. So, Table 8-2 explains the key similarities and differences.

TABLE 8-2   **Behavior by Group Creation Method**

Feature	Outlook First	Teams First
Comment availability	Comments enabled immediately	Comments disabled until mailbox visibility is enabled
Comment storage	Group mailbox enabled immediately	Group mailbox enabled after visibility is enabled
Comment access	Outlook conversations and Planner	Outlook conversations and Planner
Teams integration	Comments based in Outlook	Comments based in Outlook

Create the group in Outlook if the team will mainly collaborate by email. Create the group in Teams if the team mainly collaborates by Teams chat.

It may seem like you have to do a lot just to enable comments in Planner based on the information in this chapter, but it's simple. When you understand groups, where comments are stored, and how you and the team access those comments, it's just a matter of creating the group in whichever way makes the most sense for how your team collaborates best.

Linking a plan to a group

After you've created a group, the first step to using collaboration features is to attach the plan to a group. The process is the same whether you're creating the plan from the Planner web app or from Teams. The following example uses the Planner web app:

1. **Open Planner and click New Plan at the bottom of the navigation pane.**

    The Create New window appears.

2. **In this example, click Simple Plan.**

    The Templates window appears.

3. **Click Use Template.**

    The Create a Plan from a Template window appears.

4. **In the Name field, enter the name for your plan.**

5. **From the Add to a Group (optional) drop-down, select the existing group to which you want to attach the plan (see Figure 8-9).**

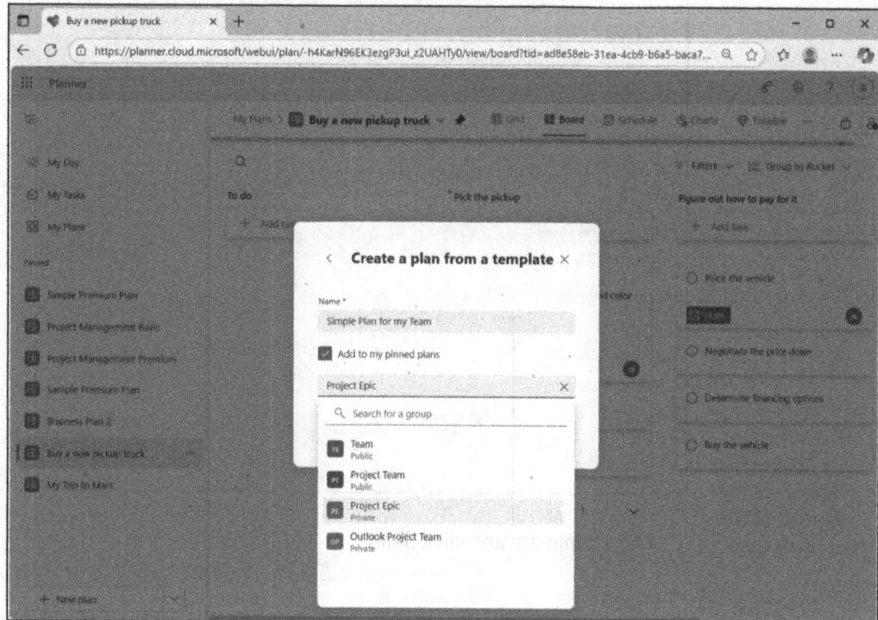

FIGURE 8-9: Choose an existing group for the plan.

Adding and viewing comments

With your group created and a plan attached to the group, you're ready to start using comments. Creating them is easy. Just follow these steps:

1. **Open any group-linked plan.**

2. **Open a task's Details page.**

3. **On the Details page, scroll down to find the Comments section (see Figure 8-10).**

4. **Click in the Comments field and type your comment.**

5. **Click Send.**

   The comment appears at the top of the comments list just under the Comments field.

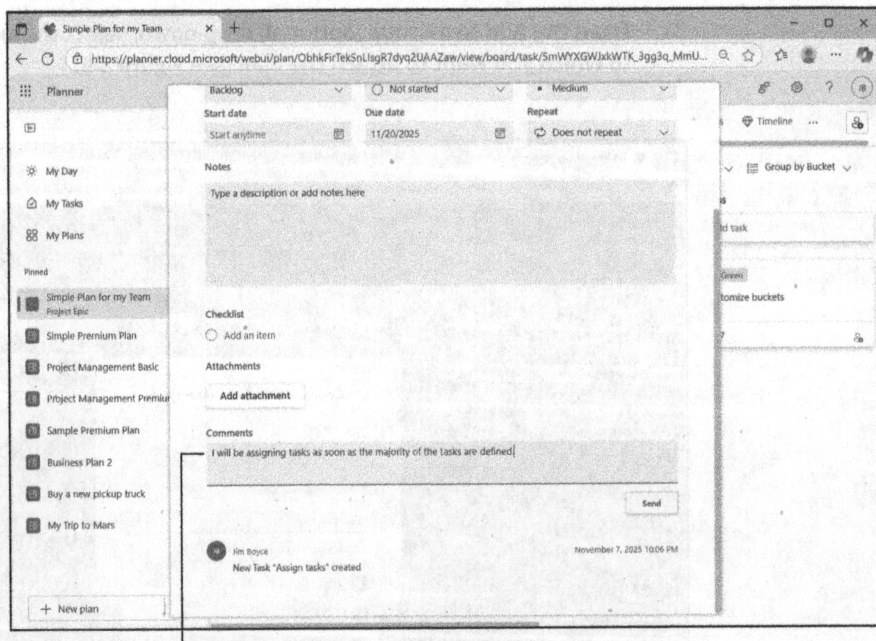

FIGURE 8-10: Comments appear at the bottom of the task Details page.

Comments appear on a task's Details page regardless of whether you're working with the plan in the Planner web app or in Teams.

WARNING

You can't edit or delete a comment after it's created. Like bad choices in life, Planner comments live forever. So, take extra care when adding comments to tasks to ensure that they not only use good grammar and spelling, but will also be interpreted as you intend by others who will see the comment.

Now, let's assume that you're working with an Outlook-integrated plan and you've added a comment to a task. Not only will that comment show up in the Planner web app and in the Teams Planner app, but it also will show up as an email in Outlook. Figure 8-11 shows an example.

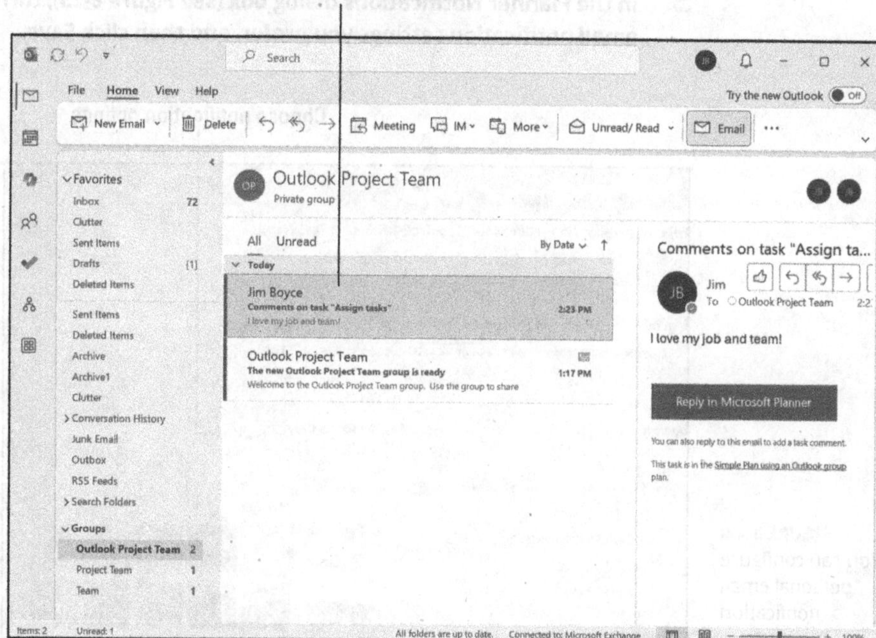

FIGURE 8-11: Comments appear in Outlook as emails, as well as in Planner.

Setting Notifications

Planner offers several notification features to keep you informed about task updates and other activities in a plan, including task assignment, due date reminders, and comments. These notifications appear in Outlook or Teams depending on how your plan has been created. For Outlook-integrated plans, notifications appear in Outlook; for Teams-integrated plans, they appear in Teams. Plans that are created outside of Teams but linked to the group will also appear in the Teams Activity feed.

Planner notifications rely on the email capabilities of Microsoft 365 Groups, so, by default, there is no setup or configuration you need to do to use notifications in Planner. Notifications will work for any plan that is linked to a group. You can, however, change a few settings to tailor how notifications work for you.

Follow these steps to configure your personal email settings:

1. **Open the Planner web app and click the gear icon in the upper right of the window.**

2. **Click Notifications (see Figure 8-12).**

3. **In the Planner Notifications dialog box (see Figure 8-13), turn on or off the email notification settings you prefer, and then click Save.**

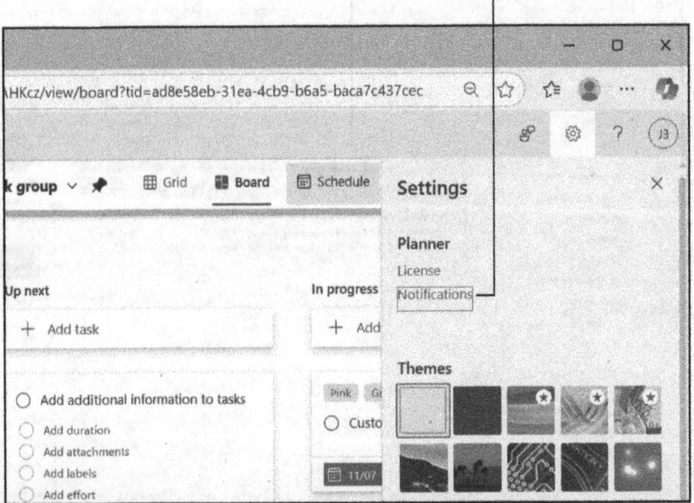

FIGURE 8-12: You can configure personal email notification settings.

With Teams-enabled plans, you receive notifications in the Teams Activity feed when someone assigns a task to you. If you have pop-up banners enabled, you'll see notifications there for task assignment. Even for Teams-enabled plans, Planner still sends email notifications to you for task assignment, due date reminders, and task comments.

Planner sends you a comment notification only if you've previously commented on the task.

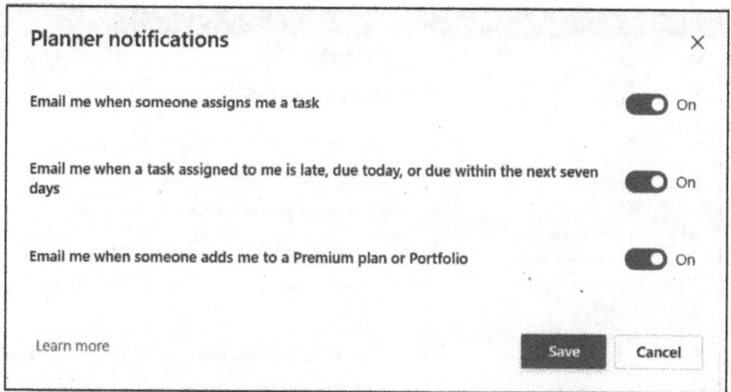

FIGURE 8-13: Use the Planner Notifications dialog box to configure email preferences.

Follow these steps to configure your Teams notification settings:

1. **Open Teams.**

2. **Click the ellipsis beside your account icon in the upper-right corner.**

3. **Click Settings (see Figure 8-14).**

4. **Click Notifications and Activity.**

   The Notifications and Activity pane appears (see Figure 8-15).

5. **Configure notifications to suit your preferences.**

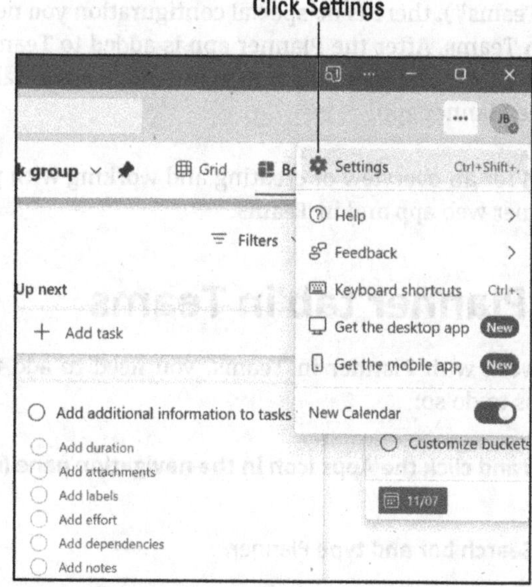

FIGURE 8-14: Open Teams Settings to configure notification options.

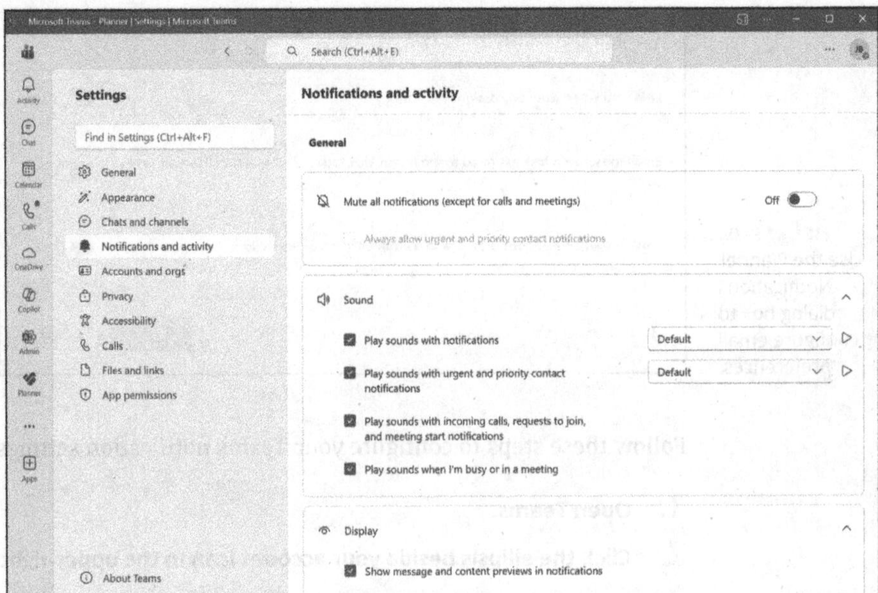

FIGURE 8-15:
The Notification and Activity pane in Teams.

Integrating with Teams

The Planner app for Teams is the tool that enables you to create and work with plans through Teams. Other than adding the app to Teams (explained in "Adding a Planner tab in Teams"), there is no special configuration you need to perform to enable Planner in Teams. After the Planner app is added to Teams, you can work with plans, tasks, views, and other Planner elements in essentially the same way as you do with the Planner app.

TIP

Turn to Chapter 5 for an overview of creating and working with plans, tasks, and views in the Planner web app and in Teams.

Adding a Planner tab in Teams

Before you can work with Planner in Teams, you need to add the Planner app. Follow these steps to do so:

1. **Open Teams and click the Apps icon in the navigation pane (see Figure 8-16).**

2. **Click in the Search bar and type** Planner.

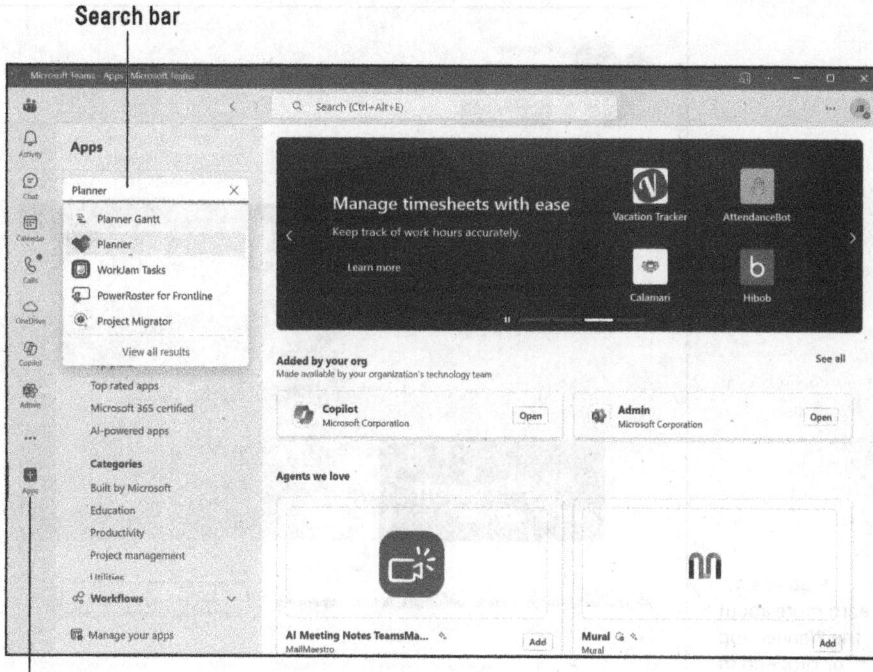

FIGURE 8-16:
You can add
other apps to
Teams, including
Planner.

3. **Click the Planner app in the resulting list.**

The Planner dialog box appears.

4. **Click Add (see Figure 8-17).**

Teams displays a Success dialog box.

5. **Click Open to open Planner in Teams.**

Pinning the Planner app

Adding the Planner app to Teams doesn't automatically pin the app to the navigation pane. Pinning the app to the navigation pane lets you quickly open Planner. Here's how to pin it:

1. **In Teams, click the ellipsis near the bottom of the navigation pane list.**

2. **Locate the Planner app in the resulting list and click it.**

The pinned Planner app now appears in the navigation pane.

3. **Right-click the Planner app and click Pin.**

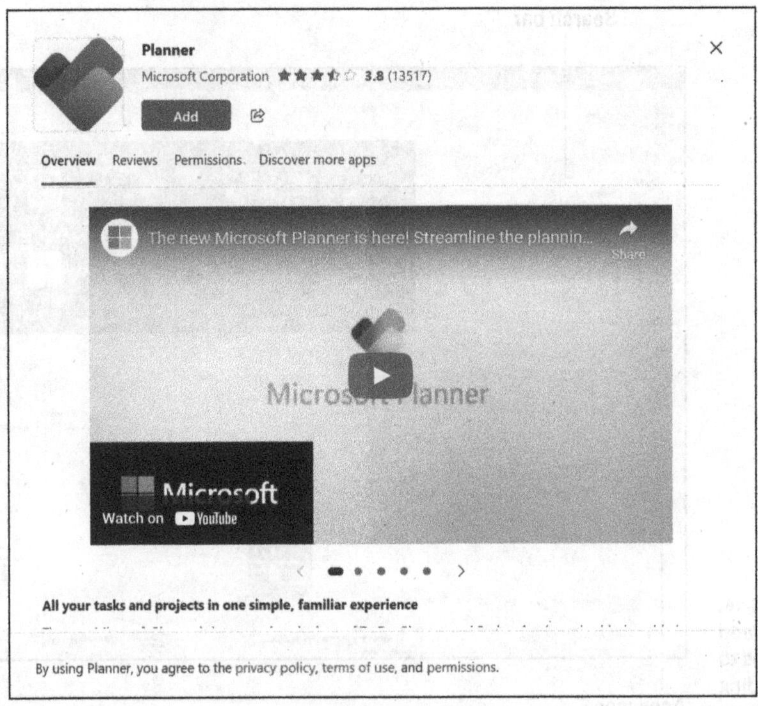

FIGURE 8-17: Learn more about the Planner app or click Add to add it to Teams.

If you don't use Planner often or just want to streamline your navigation pane, you can easily unpin the Planner app. Just right-click the Planner icon in the navigation pane and click Unpin.

TIP

If for some reason you stop using Planner in Teams (for example, you decide to work exclusively with the Planner web app), you can uninstall the Planner app from Teams. To do so, right-click the Planner icon in the Teams navigation pane and click Uninstall.

Using Planner with SharePoint

Each Microsoft 365 group includes a SharePoint Online site that includes a document library for storing and sharing documents. The site also includes a shared OneNote notebook that you can use to share notes, create an informal wiki, and store other information shared by the group. Because Planner is also a shared component for the group, you can integrate Planner and SharePoint to collaborate with others in the following ways:

>> **Sharing documents:** You can use the SharePoint site's document library to store project documents and link them to tasks for easy access to the file from the task. You can also upload documents to SharePoint when creating a task.

>> **Sharing a notebook:** OneNote is a free-form notebook, meaning you can create whatever section and page structure you need for a given scenario. You can use OneNote to keep meeting notes, create an FAQs page or wiki, and store other information in OneNote.

>> **Using SharePoint pages and lists:** You can create pages and lists in SharePoint related to the team and/or projects. You can include links in a task that direct people to a document, page, or other item stored in SharePoint.

>> **Embedding Planner in SharePoint pages:** The Planner web part for SharePoint enables you to embed Planner content directly in SharePoint pages, enabling your team to work with tasks directly from the SharePoint page.

There are lots of reasons to use SharePoint in conjunction with Planner. SharePoint not only provides the means for document sharing and the other capabilities I describe earlier, but also adds capabilities for access control, file versioning, search, and compliance.

Using SharePoint and Planner together goes beyond the scope of this chapter, but I mention it here as an important means of collaborating with your team on your plans and projects. For a deeper discussion of SharePoint integration, check out Chapter 13.

Integrating Planner with Outlook

I describe the process and reasons for enabling Outlook integration earlier in this chapter in the section, "Creating a group from Outlook." Comments and notifications are a key reason to use Planner and Outlook together, but the two apps offer other benefits as well. For example, you can use the group's shared calendar for meetings and events related to your team's projects. The following sections explore this scenario and others for integrating Planner and Outlook.

Using a shared calendar

Team projects don't happen in a vacuum. Your teammates don't work on their tasks in the solitude of a dusty, candlelit garret. Every project includes meetings and events: a project kickoff meeting, status or scrum meetings, one-off meetings to discuss specific issues or challenges, and so on. Meetings can consume as much as 25 percent or more of a project's actual time, so whatever you can do simplify and streamline those meetings is critical to project success, team morale, and productivity.

Plans that are linked to a Microsoft 365 group can take advantage of the group's shared calendar. However, tasks don't automatically show up on the shared calendar. In most plans, having the tasks on the calendar would result in an unmanageable calendar due to the sheer number of tasks in the plan, particularly when sharing calendar space with all the team's meetings. Instead, you should use Schedule view and Timeline view in Planner to view tasks on a timeline.

It can make sense to add some of the very high-level or important tasks to the shared calendar. These tasks include a kickoff meeting, monthly status review, product launch, or other key items that you've added as tasks in the plan. In these situations, create a calendar item and link the task into it:

1. **Open Planner and locate the task you want linked to the calendar item.**

2. **Click the ellipsis at the upper right of the task and choose Copy Link to Task (see Figure 8-18).**

3. **Open Outlook and then open the shared group calendar (see Figure 8-19).**

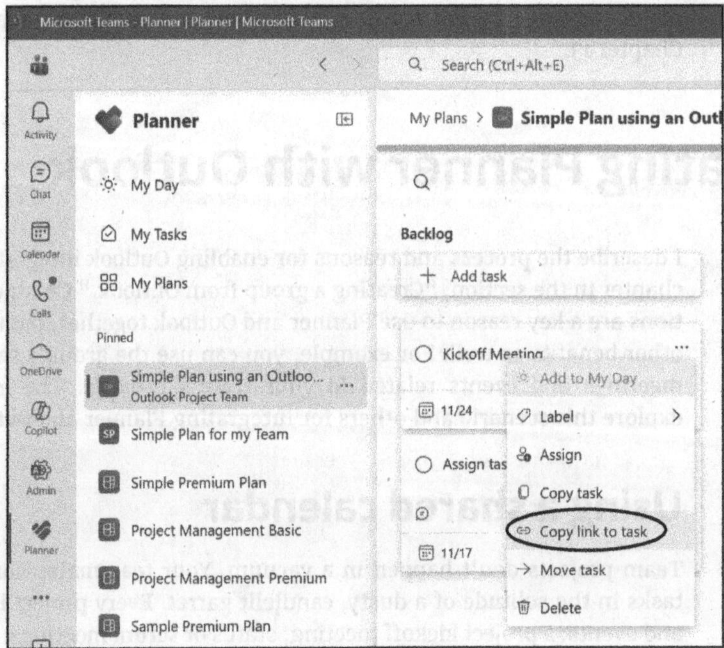

FIGURE 8-18:
Copy the link to the task.

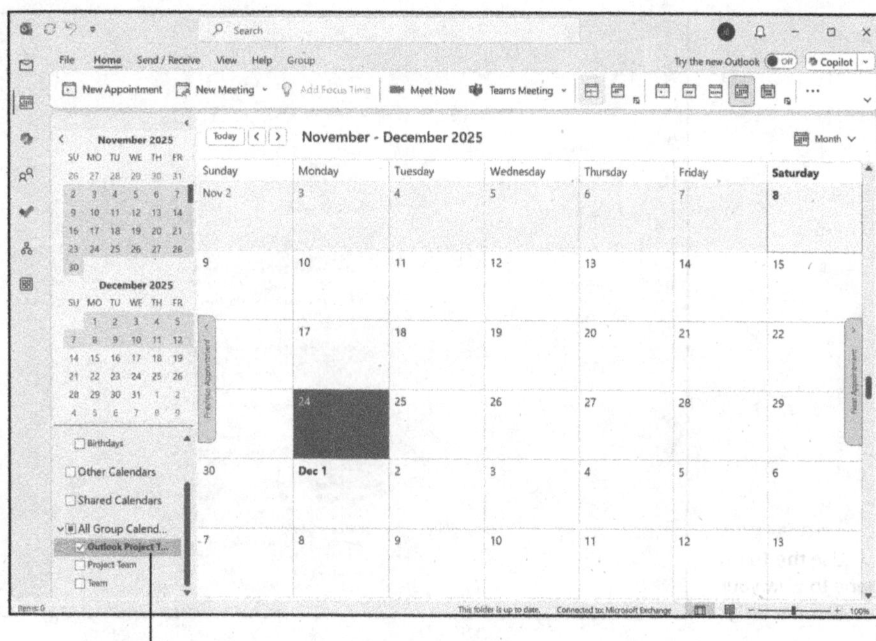

FIGURE 8-19: Open the shared group calendar and create the meeting item in Outlook.

4. **Create a meeting invitation on the task's due date and insert the link into the body of the invitation.**

5. **Add meeting participants and click Send to send the meeting invitation.**

Adding a link to the task in the meeting invitation gives the team quick access to the task from Outlook. Team members can open the meeting in Outlook and click the task's link to open the task in Planner.

Viewing your tasks in Outlook

Outlook includes a Tasks folder that you can use to create and work with personal tasks. However, the Tasks folder and the tasks you create in it have no connection to Planner. Instead, you can use the To Do app in Outlook to view and work with your Planner tasks. Figure 8-20 shows the To Do pane in Outlook.

Personal tasks that you've created in a personal plan appear in the To Do pane in Outlook in Tasks view by default. Planner items that are assigned to you appear under the Assigned to Me view. Just as in Planner's My Day view, any of your tasks that are due on the current day appear in To Do's My Day view.

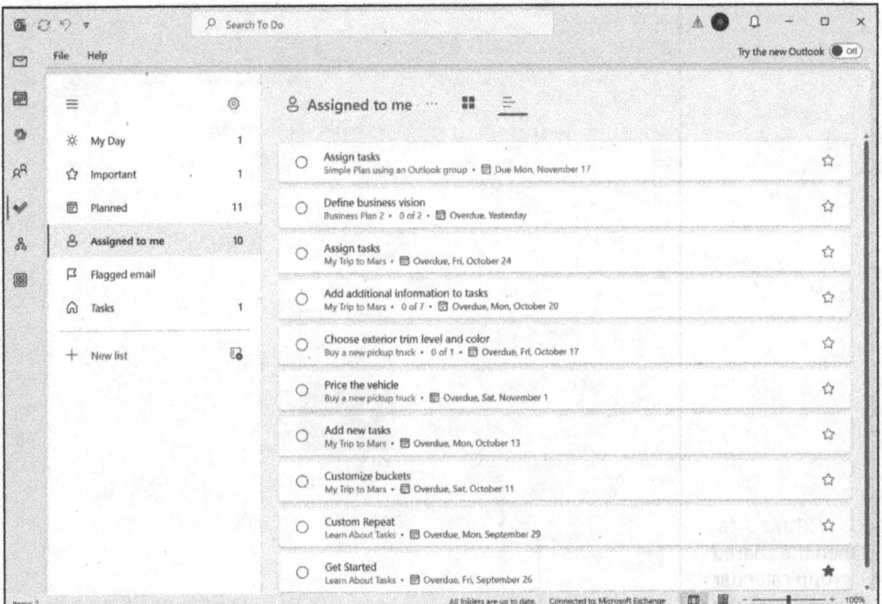

FIGURE 8-20:
Use the To Do pane to view your Planner tasks.

"Why not just work in the Planner web app instead?" you ask. Recent studies have shown that people who use Outlook spend as much as 60 percent of their time communicating through email and Teams. Most active users spend as much as 25 percent of their work week reading and writing emails. So, if you use Outlook, it's a good bet that you spend lots of time in the app. When you need to take a quick look at your tasks or mark them complete, you can do so in Outlook without having to switch to the Planner web app or Teams to do so. Likewise, if you spend a lot of time in Teams, you can view and work with your tasks directly in Teams without opening the Planner web app.

There is no right or wrong way to work with tasks — just different ways that suit your work style and the apps you use most during the day.

REMEMBER

Sharing Plans and Managing Permissions

Plans are tied to Microsoft 365 Groups, so you don't actually share a plan to give team members access. Instead, you add people to the group, which gives them access to the plan. There is no granularity to plan sharing — anyone in the group has access to all the plans that are tied to the group. There is one caveat: Users must have a license that includes Planner, such as Microsoft 365 Business,

Enterprise, or Education. So, there may be a scenario in which some users in a group don't have Planner access, which means they won't have access to the plans. Using that mechanism to manage plan access is a poor choice, however. In most cases, you won't have any insight into what license a given user is assigned, so there's no way for you to know whether they'll have access. Even if you did have that visibility, managing access through licensing introduces a level of complexity that is hard to track. For that reason, just create separate groups when you need to control access to specific projects and teams.

The section, "Meeting minimum requirements to use comments," earlier in this chapter explains how to create groups through Teams and Outlook, and explains the differences and advantages of each. Those sections also explain how to add users when you create the group. The following sections explain how to add people to a group after you've created it.

Adding group members through Outlook

The owner of a group can add and remove members from the group as needed. Assuming you're the owner of a group and you want to modify that group from Outlook, here's how:

1. **Open Outlook Web App at** `https://outlook.office.com` **and log in with the account that owns the group.**

2. **In the navigation pane, click More Apps and choose groups.**

3. **In the Groups pane (see Figure 8-21), click the group you want to manage.**

4. **Click Add Members and use the resulting dialog box to search for and add members to the group.**

You can modify other group properties through the Microsoft 365 Groups app, such as the name, description, and privacy options. Click the Edit Group button to modify the group's properties.

The Microsoft 365 Groups app gives you a place to configure your own email settings for the group. For example, if you're only managing the group for your team but you aren't part of the group's projects, you may want to limit the amount of email you get from the group. To configure your email settings for the group, open the Microsoft 365 Groups app and click the Settings (gear) icon to view and modify your email options.

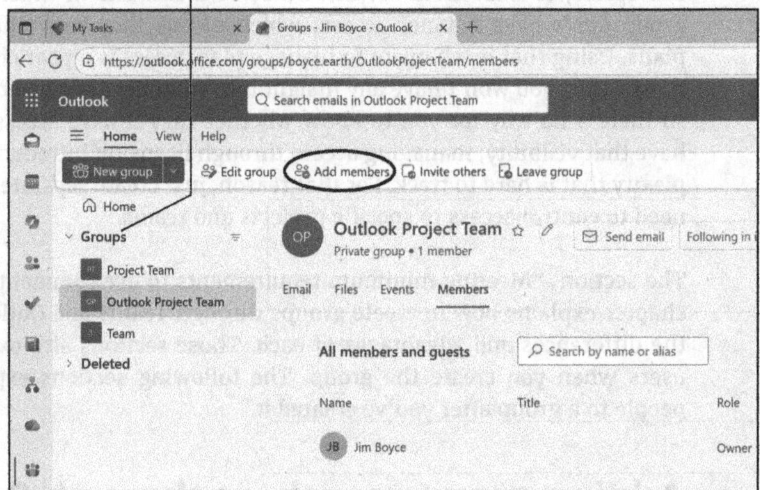

FIGURE 8-21:
Use the Microsoft 365 Groups app in Outlook Web App to manage group membership.

Depending on your license and a handful of other factors, you may also be able to use the Microsoft 365 Groups app in the classic Outlook desktop app (not in the new Outlook). If you don't see the Microsoft 365 Groups app in Outlook's navigation pane, click More Apps and choose the Microsoft 365 Groups app. If the Microsoft 365 Groups app still doesn't show up, check with your Microsoft 365 admin team to find out why — or just use Outlook Web App!

Adding group members through Teams

Because a team in Teams is tied to a Microsoft 365 group, adding someone to a team gives them access to any Planner plans that are linked to the group. (There is no granularity of control — if a person is a member of the team, they have access to all plans.) So, it's easy to give someone access to a plan in Teams:

1. **Open the Teams app and select the team you want to manage.**

   If you have trouble finding the team, open Chat and click the See All Your Teams link at the bottom of the chat pane.

2. **In the team properties pane (see Figure 8-22), click Members in the header and then click Add Members to search for and add new members to the team.**

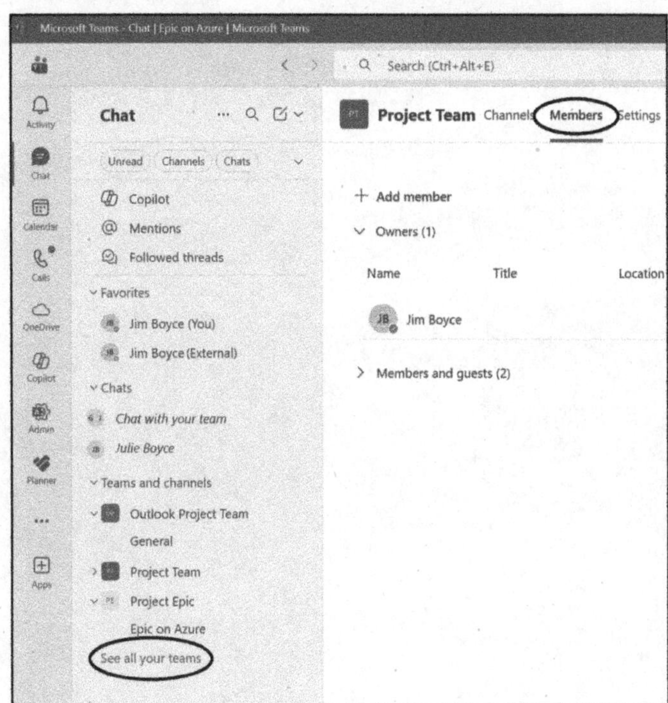

FIGURE 8-22:
Use the team's properties pane to manage membership.

Chapter 9

Managing Tasks

A good plan is the cornerstone of any project, but tasks move the project forward. Tasks *are* the project, and a tool like Microsoft Planner is the mechanism that enables you to structure those tasks into a logical sequence, break the project into manageable chunks, and distribute tasks across the doers. This all comes down to *managing task flow* with assignment, priority, and progress tracking. That's just what this chapter covers.

Managing Task Assignment

Task assignment is just what its name implies: the process of assigning a task to a person or team. The ability to assign tasks enables you to distribute the workload of a project across one or more teams. Assigning tasks to distribute workload is certainly important, but it isn't the only reason to assign tasks. Permissions, access, governance, skills, and several other factors play a part. Using the example of building a house, you would assign design tasks to the architect, engineering tasks to a structural engineer, foundation work to a concrete or masonry team, framing to the framing carpenters, wiring to an electrical contractor, and so on. Each trade accomplishes its part of the project based on skill set.

Topics like security and governance can also factor into how you assign tasks. Using an IT project as an example, multiple teams may have the skill set to accomplish a set of tasks, but only one group may have the necessary permissions in the target systems to do the work because of how the teams are organized and the governance and change control process the organization has put in place. So, although this chapter focuses on the mechanics of assigning tasks, keep in mind that lots of factors determine which individuals and teams are assigned which tasks.

Assigning tasks

Before I dive into the mechanics of assigning tasks, let's consider a couple of scenarios. First, it's obvious that task assignment is important when a group of people are collaborating on a task — you need to not only distribute the workload but also let the assignees know what tasks they need to complete. You also need to be able to track completion by those individuals and teams. But what about a personal plan where you're the only doer? Do you really need to assign tasks to yourself? In a word, *yes*.

Even if you're the only person working on a project, there are good reasons to formally assign each task to yourself:

>> **Visibility:** Tasks you've assigned to yourself show up in My Tasks in Planner, as well as in your personal To Do list and in Microsoft Outlook, so you can work with them even when you don't have Planner open.

>> **Future changes:** Assigning tasks to yourself up front can save time when someone else starts helping with the project.

>> **Reporting:** Assigning tasks to yourself enables you to use dashboards and reports to understand your personal workload.

Whether you're assigning a task to yourself or someone else, the process is the same, but there are two ways to do it. Here's the first:

1. **Open the Details page of the task.**

   In Board view, click the task name. In Grid view, click the More Options button to the right of the task name and choosing Open Details.

2. **On the Details page, click the Assign Task button (see Figure 9-1).**

3. **In the resulting pop-up window, type the name or email address of the assignee.**

   If the assignee's name already appears in the window, click it.

4. **Repeat the process for other assignees for the selected task and then close the window.**

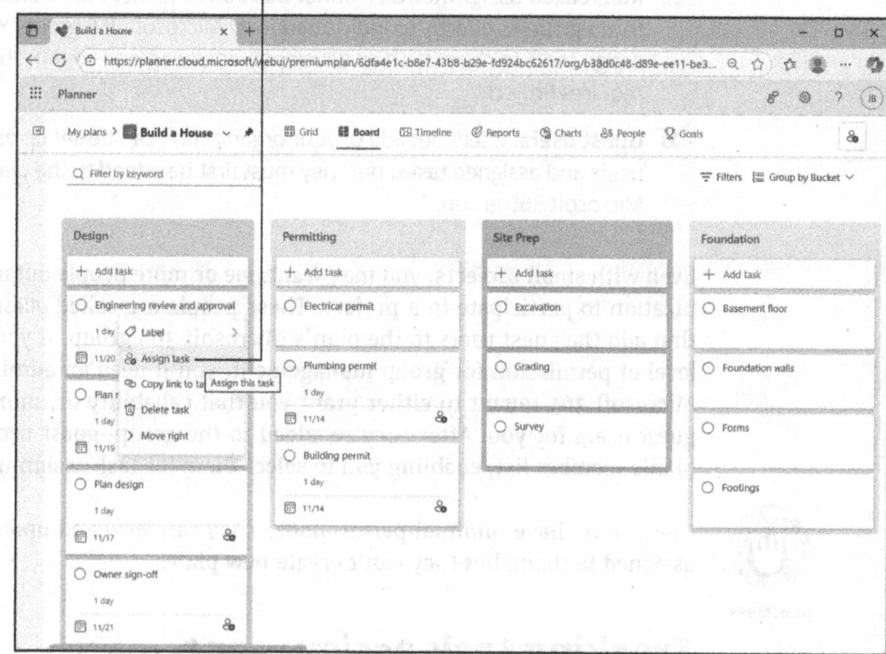

FIGURE 9-1:
Assign a task from the task's Details page.

You can also assign a task by clicking the ellipsis beside the task's name and choosing Assign. Planner displays the same people picker as on the Details page.

PREMIUM

Now that you see how easy it is to assign a task, let's dig into a slightly more difficult topic: Who or what can be assigned to a task? First, only people resources can be assigned to a task with Planner Basic and Planner Plan 1. You can't assign rooms, equipment, materials, or other non-people resources with these two versions of Planner. Instead, you need Planner and Project Plan 3 or Planner and Project Plan 5 to be able to assign non-people resources to a task. This book doesn't cover the latter two versions in detail, so this chapter describes only people assignment.

Second, there are some limitations to how you assign tasks in Planner. The following list describes these limitations:

>> **Group membership:** Users must be members of the Microsoft 365 group that is associated with the plan in order to assign tasks to them.

>> **User assignment only:** In Planner Basic and Planner Plan 1, you can only assign tasks to individuals. No version of Planner or Project supports assigning tasks to a security group.

>> **Role-based assignments:** Planner Basic and Planner Plan 1 enable you only to assign tasks directly to individuals in its Microsoft 365 group. You can't assign tasks to roles (such as developer, engineer, and so on). That capability requires Project.

>> **Guest users:** Users outside of your organization can be set up as guest users and assigned tasks, but they must first be added to the plan's Microsoft 365 group.

Even with small projects, you may want one or more people outside of your organization to participate in a project. These people are called *guest users*. You must first add the guest users to the plan's Microsoft 365 group. If you don't have that level of permission for group management, you'll need an administrator in your Microsoft 365 tenant to either grant you that capability or, more likely, add the guest users for you. After they're added to the group, guest users appear in the plan's member list, enabling you to select them for task assignment.

Guest users have minimal permissions. They can view and update tasks that are assigned to them, but they can't create new plans.

Tracking task assignment

It isn't difficult to track who's working on which tasks when you're working with only a few people on a plan. As the complexity of the plan grows, so does managing task flow, including tracking task assignments.

First are *your* tasks. You'll find those in Planner's My Tasks view. Other people's tasks have an indicator in the Board view at the bottom-right corner of the task card with the person's initials (see Figure 9-2). You can hover over the icon to view the person's name.

See who's assigned to a single task is handy, but it isn't very useful when you need to view task assignment across the plan. That's where filtering and grouping come in handy. For example, you can group the Board view using Assigned To, as shown in Figure 9-3. Each person's tasks are then grouped into individual columns.

Grid view offers another option for viewing all task assignments. You can sort the Assigned To column to display each person's tasks grouped together in the list, or you can filter the view to see only the tasks that are assigned to specific people. To filter, open Grid view, click Filters, and click the names of the people whose tasks you want to view. Figure 9-4 shows an example.

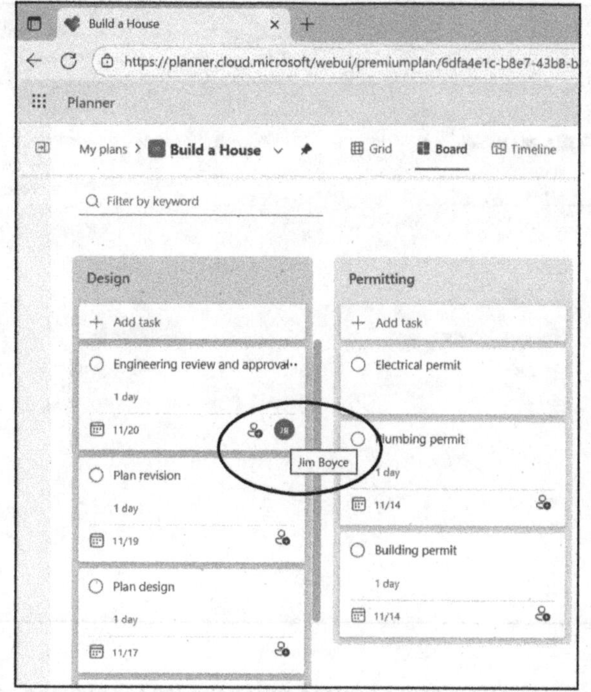

FIGURE 9-2: View task assignment in the task card of the Board view.

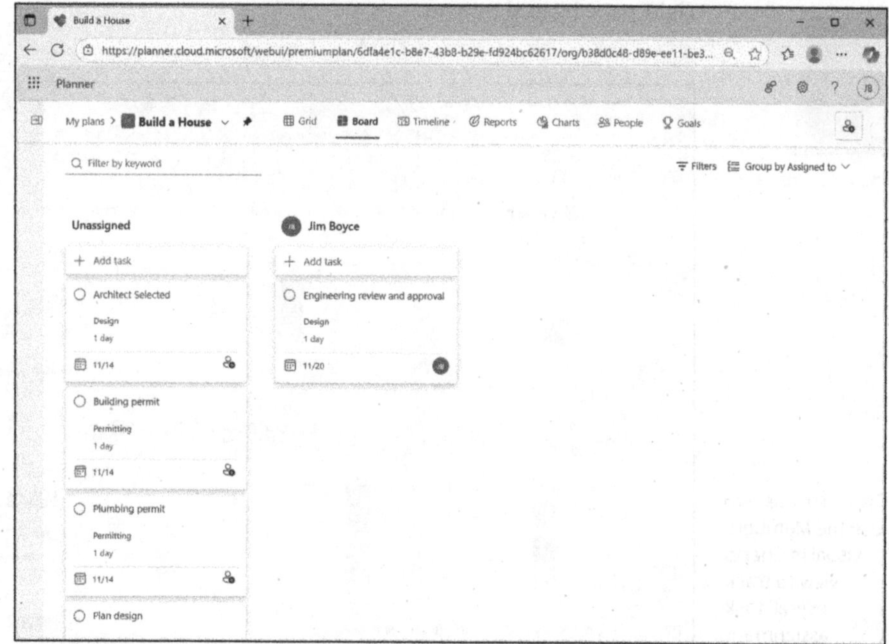

FIGURE 9-3: Group the Board view by Assigned To for viewing task assignment by person.

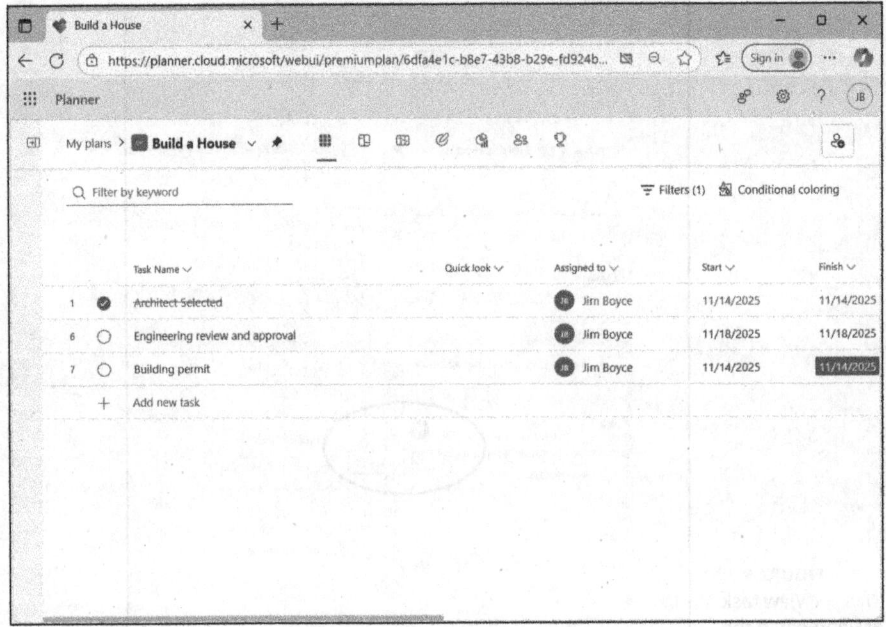

FIGURE 9-4:
Sort and filter
Grid view to view
tasks by
assignment.

Charts view also offers some limited functionality for viewing task assignments. The Members visual shown in Figure 9-5 is one example. You can also group the right pane by assignment.

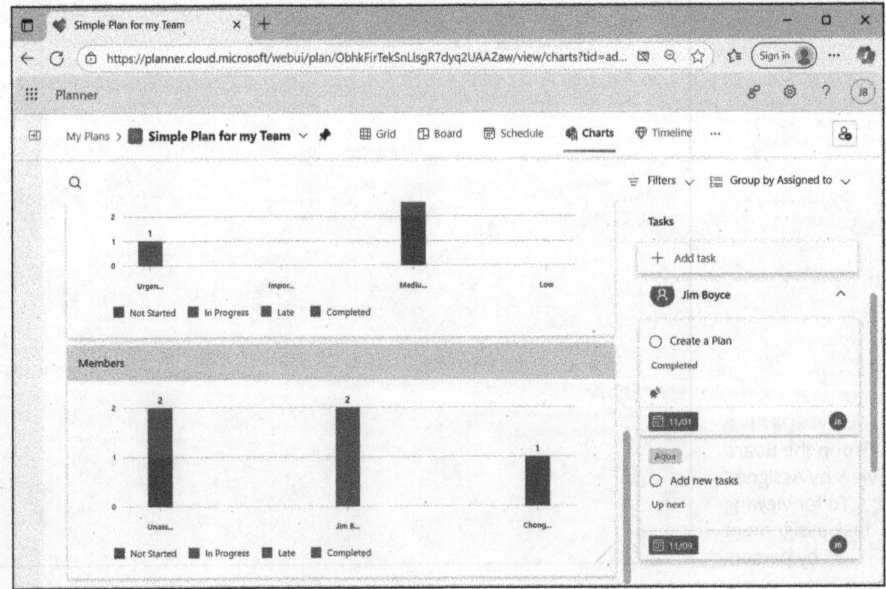

FIGURE 9-5:
Use the Members
visual in Charts
view to track
overall task
assignment
and status.

The features for viewing assignment that I've covered to this point in the chapter are all available with Planner Basic. They're certainly useful, but as you may expect, Planner Premium versions offer some additional capabilities. When you open a Premium plan, Planner Plan 1 gives you some additional ways to view task assignments. For example, the Effort per Person visual in Charts view shows total estimated effort for all tasks assigned to each person, and you can filter the visual using any of several options in the Filter Tasks pane, as shown in Figure 9-6. This visual is helpful when you need to examine resource loading to see who may be overcommitted and who on the team may be able to pick up additional tasks.

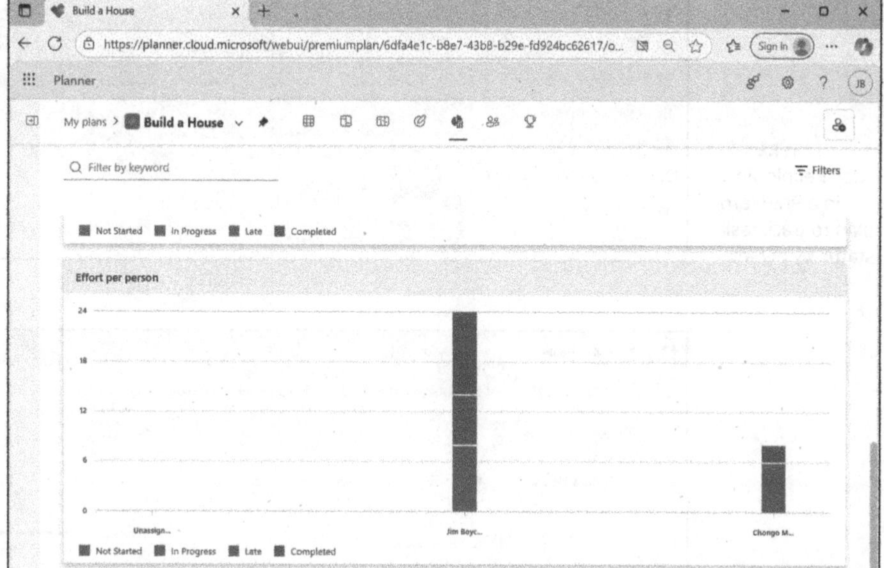

FIGURE 9-6: Use the Effort per Person visual in a Premium plan to view task load.

In a Premium plan, People view is the next place to go to view task assignments. The view rolls up all tasks by assignment and enables you to visualize assignments and completion in different ways. For example, Figure 9-7 shows the statistics for a Premium plan with cards displaying the number of completed, incomplete, and late tasks for each person. You can group the view by bucket, progress, and other properties to tailor the view.

Finally, Timeline view is useful when you want to see resource allocation for tasks across a Gantt chart. Tasks in Timeline view show a person icon beside each task to indicate ownership, and you can filter the view to show tasks on the timeline that are assigned to specific individuals so you can see where their efforts are required throughout the plan. In Figure 9-8, I've filtered Timeline view to show only the tasks assigned to me.

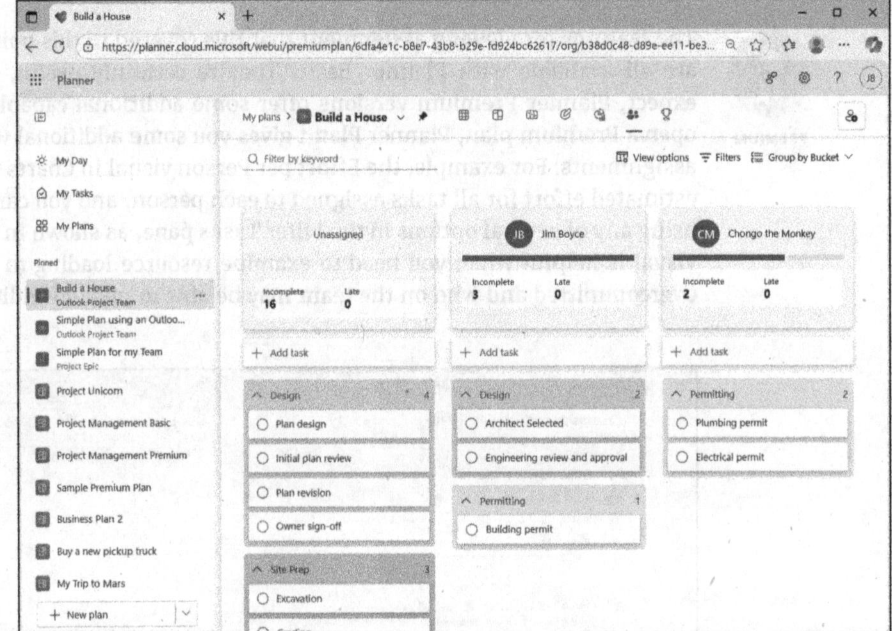

FIGURE 9-7:
Use People view in a Premium plan to track task status by person.

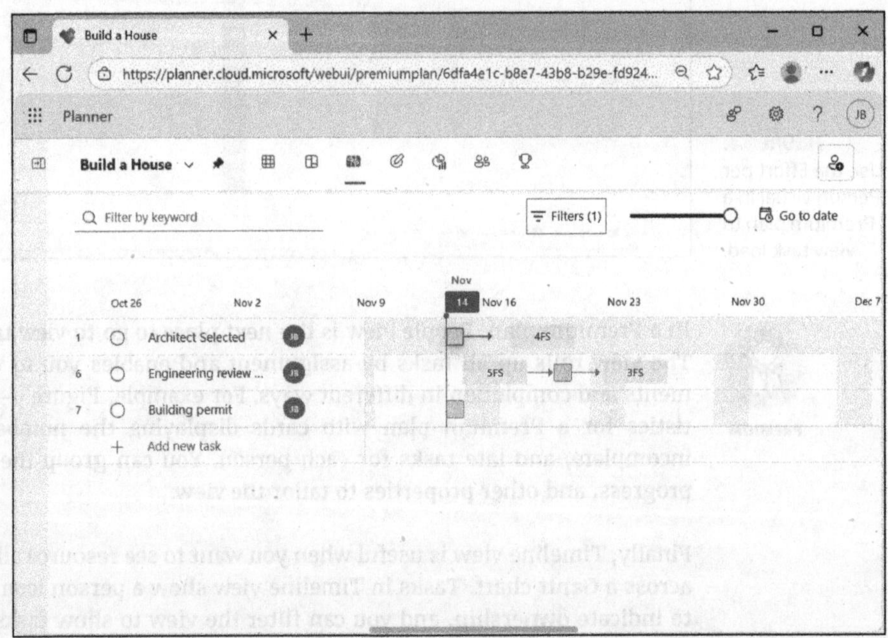

FIGURE 9-8:
Use the Assigned To filter in Timeline view to visualize effort across time for individuals.

A Gantt chart is a visual representation of the timeline of a project. It maps the tasks across a visual timeline, such as a horizontal bar representing when each task starts and is due. The Gantt chart also shows the relationships of tasks, including overlapping tasks and dependencies. Gantt charts are a powerful tool for visualizing the flow of tasks within a project.

Planner and Project Plan 3 and Planner and Project Plan 5 add more capabilities for not only assigning tasks but also visualizing task assignment and workload. These features include visualizing resources and their assigned tasks in a table, resource names and tasks on a Gantt chart, grouping by assigned resource, cross-project resource analysis, and several advanced views for visualizing and reporting on assignments and resource utilization.

Copying and Moving Tasks

The amount of flexibility for change in a plan depends on several factors, but flexibility decreases the more you work through a project. At the initial planning stage, flexibility is high. In the closing phase, flexibility is minimal. Other factors that influence flexibility include project complexity, risk level, stakeholder agreement, and contractual obligations. The higher the risk level, for example, the less flexibility you'll likely have for change. In all cases, best practices dictate implementing a change control process that defines change thresholds, approvals, communication, and other elements.

Change takes many forms: rearranging tasks, modifying requirements, reassigning tasks, changing scope, adjusting costs, and so on. Regardless of the stage of a plan, there are situations when you need to move tasks around in the plan. This certainly happens a lot in the early stages of a plan when you're fleshing out the tasks and deciding where they should fall in the overall timeline and which individuals or teams should own each task.

Planner gives you the capability to easily move tasks around within a plan, as well as copy them to other plans. The former lets you move tasks between buckets for example, which can be a common action when you're first setting up the tasks for a plan. But why the latter? Why would you want to move tasks from one plan to another? Here are some examples:

>> **Project transitions:** As a project transitions between stages, it's often necessary to move tasks from the initial plan to subsequent plans. For example, as you transition from implementation to steady-state support of a project, the support-stage tasks can move to the support team's plan. Or, the tasks may simply move into other buckets in the same plan.

>> **Segmenting a large plan into smaller plans:** Complex projects — particularly those that encompass multiple teams — can often be segmented into multiple smaller plans. In this scenario, you can copy tasks from the primary plan to the secondary plans.

>> **Consolidating plans:** Sometimes you need to segment a larger plan into multiple plans, but there also are times when you need to do the reverse: Consolidate smaller plans into an overall plan.

>> **Realigning tasks:** During a project, a group of tasks may need to move between teams or responsible parties. In that scenario, you can copy or move the tasks with the plan or into other plans.

>> **Recurring projects:** Sometimes you'll use an existing plan to create a new one, and the ability to copy tasks from the current plan to the new one can significantly speed up the process of creating the new plan and brings consistency between plans.

>> **Changing priorities:** Plans are not written in stone and changes in priorities happen. In those situations, you may need to copy or move tasks between buckets in a plan or across plans.

There are differences in copying tasks with Basic and Premium plans. In Basic plans, you can copy and move tasks within a plan or to other plans. In Premium plans, you can copy and paste tasks within the same plan but not to other plans. Likewise, with Premium plans you can move tasks within a plan but not to another plan.

The differences in behavior between Basic and Premium plans for copying and moving tasks will likely change over time as the Planner development team continues to develop features for both.

Regardless of the reason you need to move or copy a task, the process is simple. Use the following steps to copy a task in a Basic plan:

1. **Locate the task that you need to copy.**

You can use either Grid view or Bucket view.

2. **Click the ellipsis beside the task name and choose Copy Task (see Figure 9-9).**

The Copy Task dialog box appears (see Figure 9-10).

3. **Type a new name for the task if needed.**

4. **Choose the target plan and bucket where the task will go.**

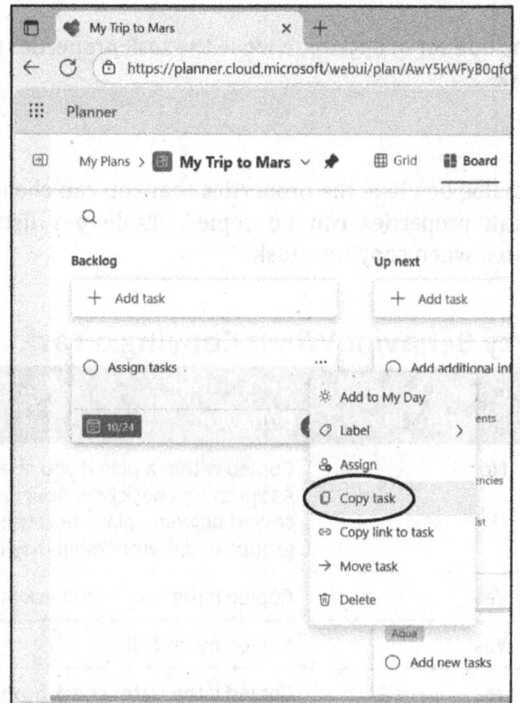

FIGURE 9-9:
Choose the **Copy Task** command in the task's context menu.

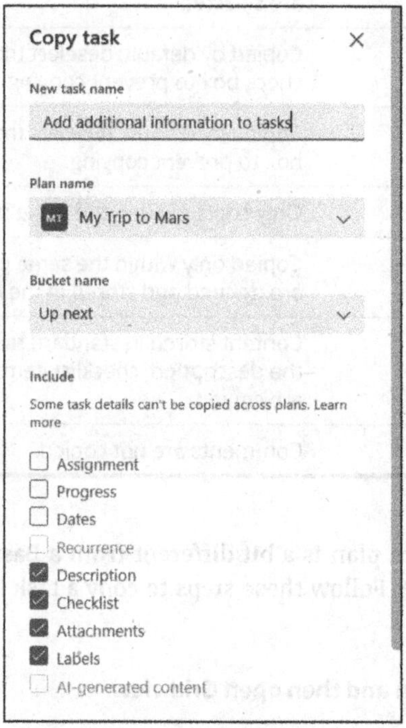

FIGURE 9-10:
Use the Copy Task dialog box to specify the copy parameters.

5. **From the Include set of options, choose the task properties to be copied with the task.**

6. **Click Copy.**

The Copy Task dialog box lists the properties that you can choose to copy for the task. Not all task properties can be copied. Table 9-1 lists properties and how they're treated when copying a task.

TABLE 9-1   **Task Property Behavior When Copying a Task**

Property	Copied within a Plan	Copied to a Different Plan	Behavior
Assignment	Yes	No	Copied within a plan if you select the Assignment check box. Assignments are not copied between plans because the associated groups and membership may be different.
Progress	Yes	Yes	Copied if the Progress check box is selected.
Priority	Yes	Yes	Copied by default.
Dates	Yes	Yes	Copied if the Dates check box is selected.
Recurrence	Yes	Yes	Copied if the Dates and Recurrence check boxes are selected.
Descriptions	Yes	Yes	Copied by default; deselect the Description check box to prevent copying.
Checklists	Yes	Yes	Copied by default; deselect the Checklist check box to prevent copying.
Attachments	Yes	No	Only copied within the same plan.
Labels	Yes	No	Copied only within the same plan because labels are defined and stored in the plan.
AI-generated content	Yes	Yes	Content stored in standard task fields including the description, checklist items, and attachments are copied.
Comments	No	No	Comments are not copied.

Copying a task in a Premium plan is a bit different from a Basic plan and uses a copy-and-paste mechanism. Follow these steps to copy a task in a Premium plan within the plan:

1. **Open the Premium plan and then open Grid view.**

2. **Click the ellipsis beside the task name and choose Copy Task.**

   Planner highlights the task to indicate that it has been copied.

3. **Locate the position in the plan where you want to copy the task, and click the ellipsis of the task immediately following that location.**

4. **Choose Paste Task.**

   Planner pastes the task above the selected task using the same name as the original.

5. **Rename the copied task if needed.**

The difference in process between Basic and Premium plans makes sense when you consider the result. Premium plans use a copy-and-paste mechanism because you're copying the task within the same plan, just like copying a sentence from one place to another in a Microsoft Word document or copying contents from one cell to another in Microsoft Excel. Basic plans give you the additional option of copying to another plan, so the interface reflects that difference.

When you copy and paste a task within a Premium plan, you may notice the Insert Task Above command in the task's context menu. Use that command when you want to insert a new task between two others. Planner renumbers the tasks accordingly.

Moving a task is similar to copying it. When moving a task in a Basic plan, you choose the destination plan and bucket. The destination plan and bucket can be the same as the current plan or they can be different. If you choose the same plan and bucket, Planner creates a new task in the same bucket with the same name (although you can rename it as you copy it by entering a new name in the Copy Task dialog box).

Follow these steps to move a task in a Basic plan:

1. **Open Planner and locate the task you want to move.**

2. **Click the ellipsis beside the task name and choose Move Task.**

   The Move Task dialog box appears (see Figure 9-11).

3. **From the Plan Name drop-down list, choose the target plan.**

4. **From the Bucket Name drop-down list, choose the target bucket name.**

5. **Click Move.**

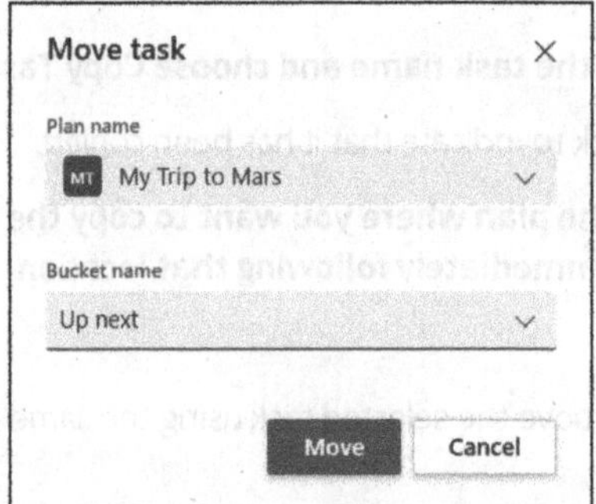

FIGURE 9-11:
The Move Task
dialog box.

TIP

Moving a task in a Premium plan changes only the task order based on the task ID. To move a task, open Grid view, hover the mouse over the six dots (the drag handle) at the very left side of the view, and just drag the task where you want it to go.

In addition to copying tasks, you can also copy an entire plan. That's a great way to create your own customized templates for specific types of projects. See Chapter 11 for more on copying plans.

Prioritizing Tasks and Managing Progress

Some tasks are critical to the successful completion of a project; others are less critical. Regardless, every task has some level of logical priority within the scope of the plan that contains it. A key action when you're building a plan in Planner is to set and manage the priority of the tasks.

Setting and visualizing task priority

Planner offers four priority levels: Low, Medium, Important, and Urgent. How you use these levels depends on a variety of factors, but it's a good idea to come to agreement with your project team on the expectations for each priority level. Defining these levels early and sharing your definitions with the team can be critical to project success. Document priority levels and corresponding expectations, process, escalation paths, and so on and store this information within the plan's document library.

As with most actions in Planner, setting task priority is easy:

1. **Open the task's Details page.**

2. **From the Priority drop-down list, choose the desired priority (see Figure 9-12).**

3. **Make any other changes to the task as needed, and then close the Details page.**

FIGURE 9-12: Choose a priority from the drop-down menu.

"Wait, I don't like those priority descriptions," you say. Unfortunately, you're stuck with them — Planner doesn't enable you to create custom priority values. However, you can use some workarounds to tailor how you set and visualize task priority. For example, maybe you'd like to have some additional granularity for Urgent tasks, such as Priority 0 for extremely critical (like life and safety issues), Priority 1 for very critical, and Priority 2 for urgent. One method to accomplish this is to use custom labels in conjunction with priority. Using this same example, create labels named Priority 0, Priority 1, and Priority 2 and assign them an appropriate color. Then, when you assign the Urgent priority to a task, also assign the corresponding label. Figure 9-13 shows an example.

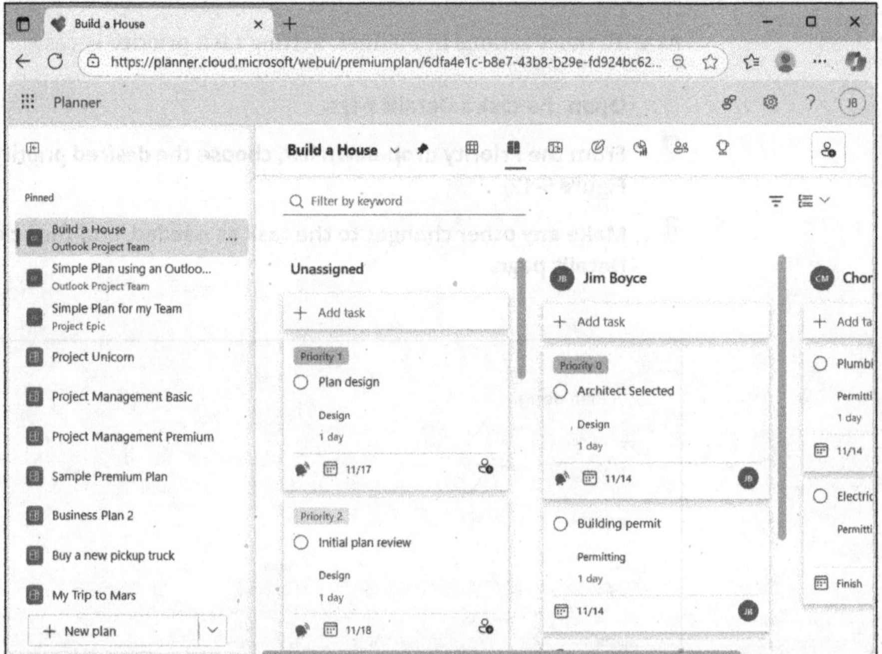

FIGURE 9-13:
Use labels in conjunction with priority to tailor priority visualization.

TIP

Keep in mind that creating labels to define sub-priorities doesn't directly tie those labels to the intended priority. Assume you create labels for Priority 0, 1, and 2 and intend to use them with Urgent priority items. This doesn't meant that you can *only* use them for Priority tasks. You can use those labels anywhere within the plan because labels and priority values have no hard linkage.

After you've set priorities for tasks, whether you're supplementing with labels or not, you can use the filtering capabilities in Planner's views to track tasks by priority. In any view, you can use the filtering and grouping capabilities to tailor the view as needed. Using the same priority/label example I describe earlier, you may filter the view by Priority=Urgent and then group by Label. Figure 9-14 shows an example. Note how only the urgent priority tasks are shown and grouped into columns by label.

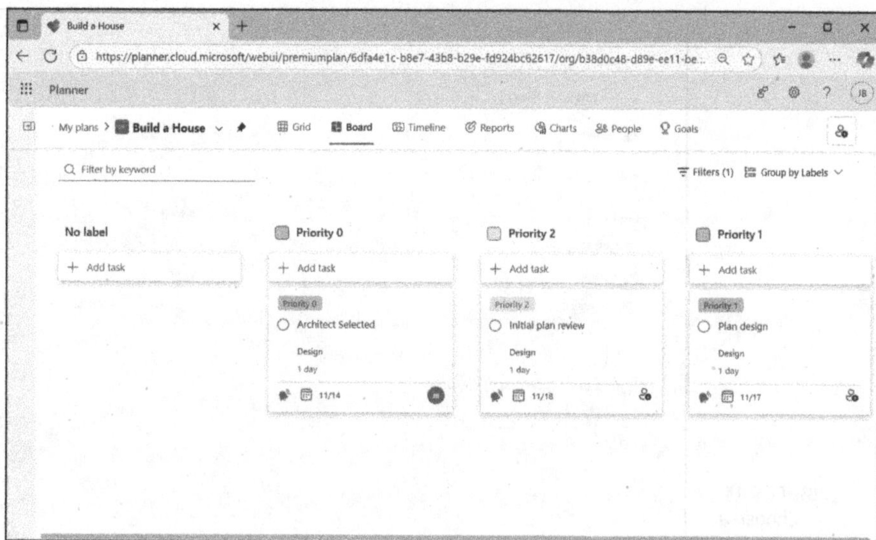

FIGURE 9-14:
Use filtering and grouping in a view to home in on specific tasks.

Managing task progress

Tasks need to progress toward completion. If it doesn't matter whether a task is finished in order to successfully complete a project, then the task may not belong in the plan. At the very least, you should mark these types of tasks as "Optional" using a label. This enables you to add items like optional features, extra testing, or other "nice-to-have" items that aren't required and track them within the project through grouping and filtering.

Basic plans

For those tasks that *do* need to be completed, you'll want to track progress throughout the project. For Basic plans you do so through the Progress property of the task. There are three predefined progress settings: Not Started, In Progress, and Completed. Here's how to assign progress:

1. **Open the Details page for the task whose progress you want to set.**

2. **From the Progress drop-down list, choose your desired progress (see Figure 9-15).**

3. **Close the Details page.**

Reporting based on progress is like reporting based on other properties like assignment, priority, dates, and so on. In whichever view you prefer, use the Filters and/or Group By controls to filter and group by Progress. Figure 9-16 shows an example of the Board view grouped by Progress.

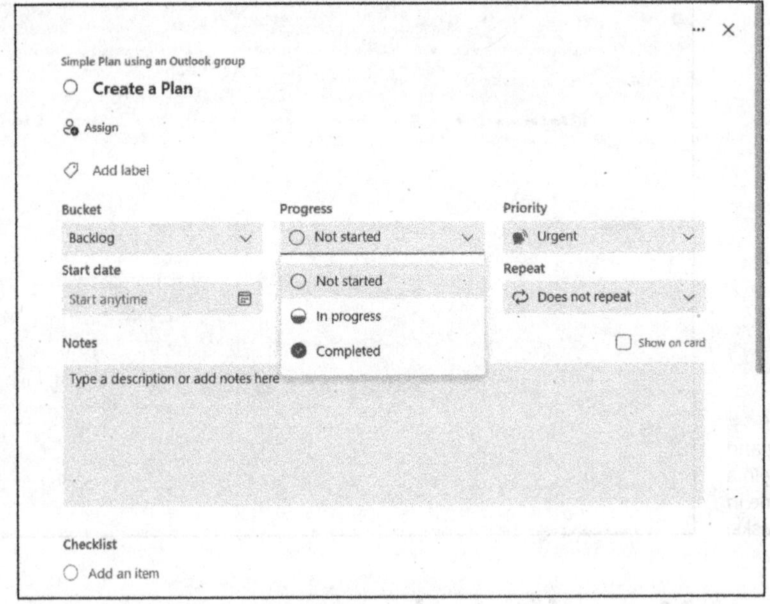

FIGURE 9-15:
Choose a progress option for a task in a Basic plan.

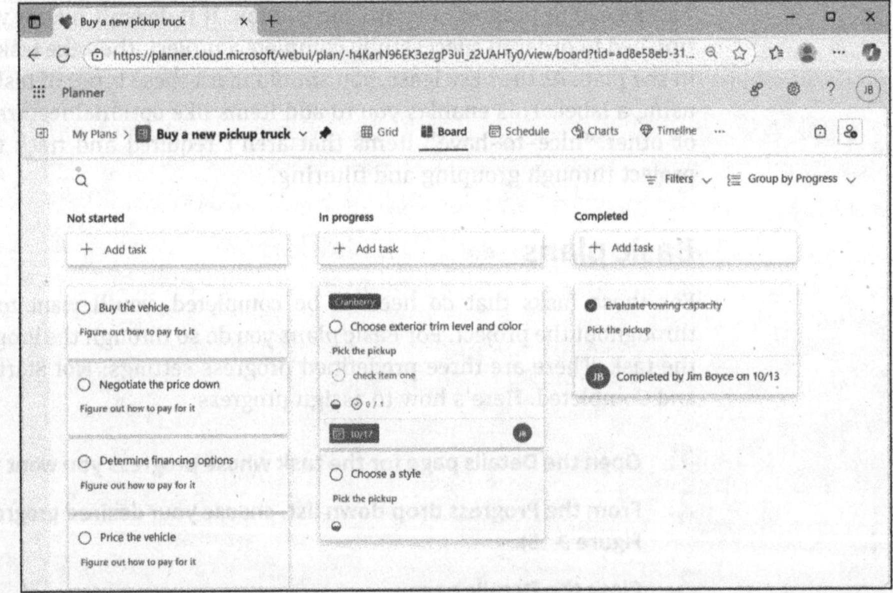

FIGURE 9-16:
Use filtering and grouping to report on tasks by progress.

As with task priority, you can't create custom progress values, but you can use labels or other mechanisms to add more granularity. Here are some recommendations:

>> **Labels:** Use labels to add more granularity to your progress categories. For tasks that are set to In Progress, for example, consider adding labels like Blocked, Reassigning, Pending Review, Ready for Testing, Waiting for Approval, and so on. Create and document these labels in the beginning of the project and ensure that all participants understand the purpose of each.

>> **Buckets:** You can create buckets to further define task progress. For example, you may want to create buckets named Blocked, Pending Review, and so on using the same logic as described in the previous bullet point. Using buckets in this way, however, requires you to move tasks between buckets as their progress changes.

>> **Notes and checklists:** You can use a similar approach with notes and checklists, setting them up as subcategories for progress. For tasks that are in progress, for example, you can create a logical flow of checklist items like Developing, In Review, Ready for Testing, In Testing, and so on. As you work through the flow of the task, check off each status as you reach it. When the items are all done, move the task to Completed.

Premium plans

Premium plans use a different method to track progress. Instead of the three progress options in a Basic plan, you use percent complete to indicate a task's progress. Tasks with 0 percent complete are classified as Not Started. Tasks with any progress greater than 0 and less than 100 percent are classified as In Progress. Setting a task's completion to 100 percent also marks the task as Completed.

Follow these steps to set progress for a task in a Premium plan:

1. **Open the task's Details page.**

2. **Enter a value in the % Complete field (see Figure 9-17).**

3. **Close the Details page.**

Just as you can with Basic plans, you can filter and group a view to show tasks by progress. The view shows Not Started, In Progress, and Completed columns based on the percent complete of each task. Figure 9-18 shows an example.

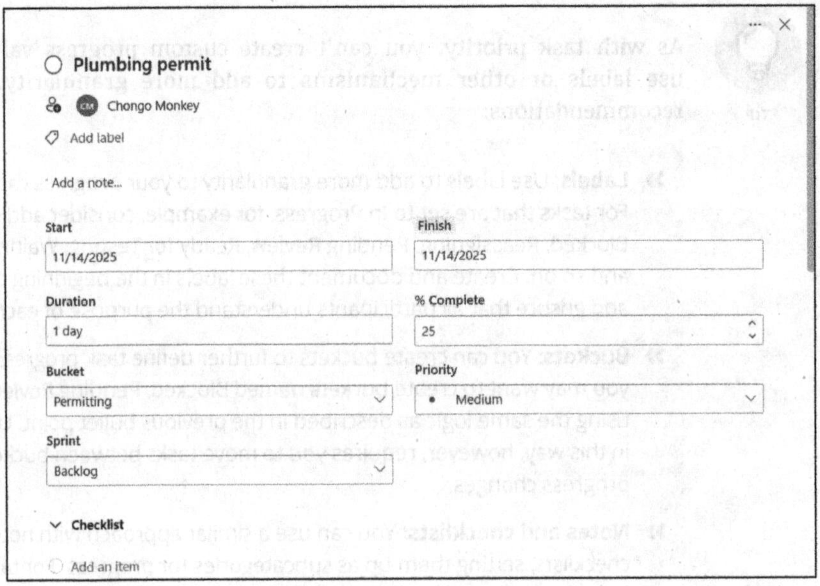

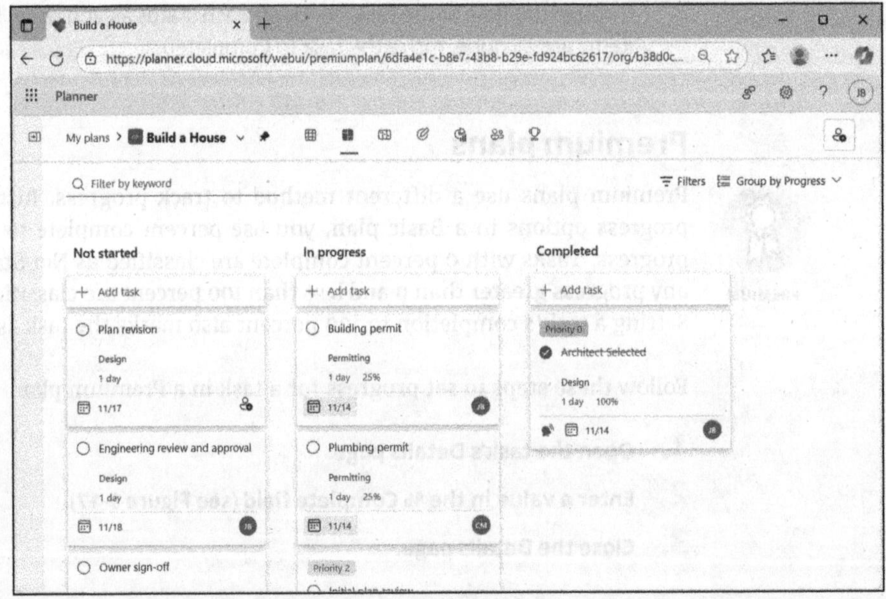

FIGURE 9-17:
Use the %
Complete field to
indicate progress
for a task in a
Premium plan.

FIGURE 9-18:
You can filter and
group a Premium
plan view
by progress.

3

Digging Deeper with Microsoft Planner

Chapter **10**

Managing Plan Structure and Flow

Building a project plan lays the foundation for the project's success. Without a good plan, most projects will either fail altogether or at least fail to meet key outcomes. But how the project flows is also important to the project's success. Sequencing tasks properly, identifying and planning task dependencies, organizing buckets effectively, and regularly checking and adjusting project flow all contribute to a well-run project. Underpinning all of that are best practices for communication, plan structure, documentation, and review. This chapter explores all these topics, starting with some best practices for structuring plans.

Implementing Best Practices for Structuring Plans

There are many ways to structure plans and manage projects, all driven by the needs of the organization and its support systems, organizational culture, team structure and work style, and not least of all, project type and scope. One size doesn't fit all, but there are common threads running through all well-structured and successful projects. The following sections in this chapter offer some overarching recommendations for project planning in general, along with specific recommendations for using Microsoft Planner to build and run projects effectively.

Whatever the mission or project, clarity of purpose and process are key to minimizing risk and ensuring success. This means that everyone needs to clearly understand the goals and objectives of the project, its requirements, individual and team responsibilities, timelines, process requirements, and workflow. Ensuring clarity in all these areas helps ensure a successful and well-run project.

Here are some tips for structuring plans:

» **Define clear goals and objectives.** Clearly document and gain consensus (and sign-off) from key stakeholders as to the goals and objectives of the project. Integrate feedback from delivery resources throughout this planning phase to ensure that risks, challenges, and alternatives are fully considered prior to finalizing the plan and beginning the project.

» **Make sure you have a clearly defined scope.** Clearly define in-scope and out-of-scope activities and outcomes, document them, and communicate them to stakeholders and doers alike. Gain approval and lock down the scope prior to the project launch.

» **Be consistent and clear in naming.** Define consistent naming conventions for buckets, tasks, labels, and other elements in the plan that clearly communicate the intent and purpose of the item. Avoid ambiguity.

» **Agree on what *progress* means.** Establish, document, and communicate a consistent understanding of how tasks should be tracked toward completion. In Basic plans, use labels to supplement the three included indicators — Not Started, In Progress, and Completed — where appropriate. In Premium plans, document effort across all tasks consistently to ensure proper progress tracking, as well as workload reporting.

» **Leave no task unassigned.** Although some tasks may remain unassigned in the very early stages of a project, focus on accurate and complete assignment early on.

» **Err on the side of providing too much detail.** As with naming conventions, include as much detail as possible when creating plan elements to ensure that all participants have clarity. Also ensure that all participants include an appropriate depth of notes and comments as they work through the plan.

» **Communicate promptly and regularly.** Meet regularly with plan participants and set the expectation that everyone should be prepared to provide status updates for their respective responsibilities. Communicate changes broadly and promptly when they occur. Set the expectation that queries should be responded to quickly and that notifications should be acknowledged. Without confirmation, there is no communication.

>> **Keep the plan focused on moving forward.** As you progress through the plan, move completed tasks to a Completed bucket in addition to marking them as completed.

>> **Use buckets effectively.** Buckets are a great mechanism for visualizing project flow, enabling you to view the plan by project phase, function, responsible team, priority, or status. Which of these attributes you choose depends on the project and several other factors, so choose a convention that fits the current plan. Use enough buckets to capture your intent, but not so many that they become difficult to visualize or manage. Use color and labels to add clarity of intent to the plan's buckets.

>> **Use granularity intentionally.** Adding granularity to a plan can bring clarity, but it can also bring confusion by adding too much detail. Using too many buckets, labels, and other elements can become overwhelming and time-consuming, so find a happy medium between too little detail and too much.

Managing Task Sequencing

A good project follows a planned sequence of interrelated tasks, with adjustments along the way to account for unforeseen issues or omissions. Without that sequence or planning, a project becomes a bit like the proverbially monkeys-and-typewriters scenario: No matter how many monkeys and typewriters you throw at a problem, you'll end up with gibberish and broken typewriters — and probably a big mess to clean up. The following sections explain how to avoid that mess.

The bulk of the work in the initial stages of creating the plan for a project is often identifying the tasks, sequencing them, and identifying relationships between the tasks. Although you can enter tasks at random into a plan — and, to some degree, that happens with many plans as you think through the project — at some point you begin to create a logical sequence of events. The following sections explore the differences for task sequencing between Basic and Premium plans.

Sequencing Basic plans

Basic plans don't use the same sequencing as Premium plans. Although each task is assigned a unique identifier, that identifier isn't visible to you. If you look at the Grid view of a Basic plan, for example, you see the task name and its other properties, but no unique numeric ID. There is also no way to change the task order within Grid view. So, the options for sequencing tasks in Basic plans are limited, but you do have a few.

The first approach is to organize the tasks within their respective buckets. Doing so is easy — just open the Board view and drag a task up or down in the list to rearrange the sequence. Continue rearranging tasks in this way until you achieve the logical sequence that you want. Keep in mind that this type of sequence is really just for visualization and has no ties to the timeline.

You can move tasks from one bucket to another by dragging the tasks, just as you can move them within a bucket.

Other options for sequencing tasks in Basic plans are much more complex. For example, you could create a Power Automate flow that automatically prepends a number to the task when it's created. You could then modify the task number, if needed. Or, use the Microsoft Graph application programming interface (API) to retrieve the metadata for the plan and tasks and then manipulate the task names in whatever way makes sense within your project framework.

Just because you *can* do these things doesn't mean you *should* do them. These approaches are complex and require some development background. Instead, if you really need automatic sequencing, consider upgrading the plan to Premium.

Sequencing Premium plans

When you enter tasks into a Premium plan, each task gets assigned a task number. By default, the task number determines the order in which the tasks appear in the Timeline view. Figure 10-1 shows an example using my Build a House plan. The left pane shows the tasks ordered by current sequence.

Sequence may not matter much in a very small plan, but it certainly becomes important as the plan complexity grows. Sequence becomes more important when you start to define task length and effort, and even more so when you start to add task dependencies. Without good sequencing, your timeline will quickly begin to look like a pile of spaghetti.

Sequencing tasks in a Premium plan falls into two separate activities:

>> **Defining sequence within each bucket:** Provides a visual means for you and others to understand the flow of tasks within each bucket. Moving tasks around in a bucket doesn't change the actual sequence based on timeline.

>> **Arranging tasks by execution sequence:** You arrange tasks in buckets for a Premium plan in the same way you do so for a Basic plan — open the Board view, choose a bucket, and start dragging tasks within the bucket to achieve the desired results.

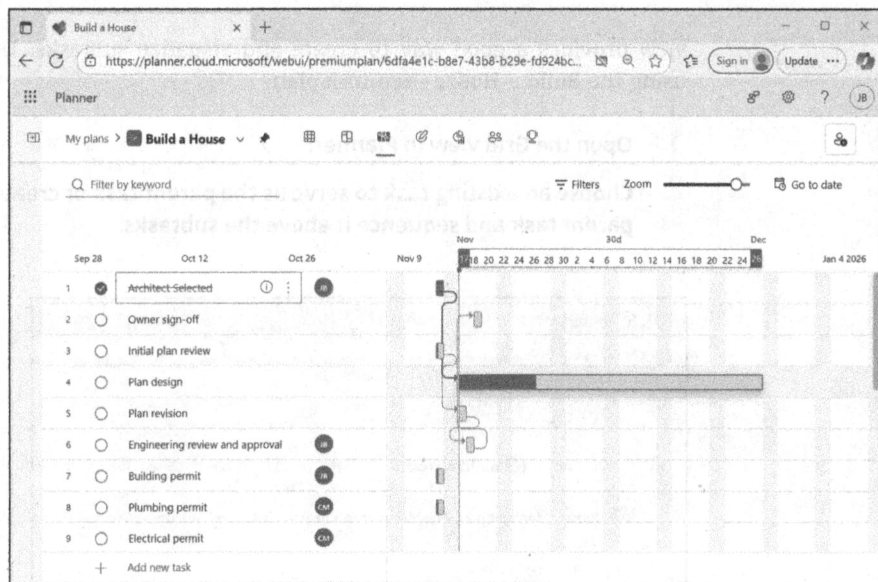

FIGURE 10-1:
Tasks are ordered by sequence in the Timeline view.

TIP

Do you *really* need to sequence tasks within the buckets of a Premium plan? After all, you're not really applying a sequence that affects the timeline. Although sequencing tasks in their buckets is optional, it's nevertheless a good way to visualize task order or even priority within the Board view. Before you start a plan, decide what's important to emphasize for the project — execution order, priority, or other property — and then organize the buckets accordingly.

When you want to sequence tasks by execution order for a Premium plan, turn to the Timeline view. Sequencing the tasks is simple. Just open the Timeline view and drag tasks into position. To do so, hover the mouse over the empty space to the left of the task number, click the six-dot icon that appears (called the *drag handle*), and then drag the task into position. Task IDs automatically renumber based on the new order.

Creating Subtasks

PREMIUM

In Planner Basic, you can use checklists to simulate *subtasks*, which are discrete activities that make up the overall task. As you complete an activity, you can check it off to indicate that it has been completed. When the entire checklist is complete, you mark the task itself as completed. In Premium plans, you have the capability to create true subtasks, each with its own assignee and other properties. The Grid view provides a good way to create and visualize subtasks. Here's an example using the Build a House Premium plan:

1. **Open the Grid view in Planner.**

2. **Choose an existing task to serve as the parent task or create a new parent task and sequence it above the subtasks.**

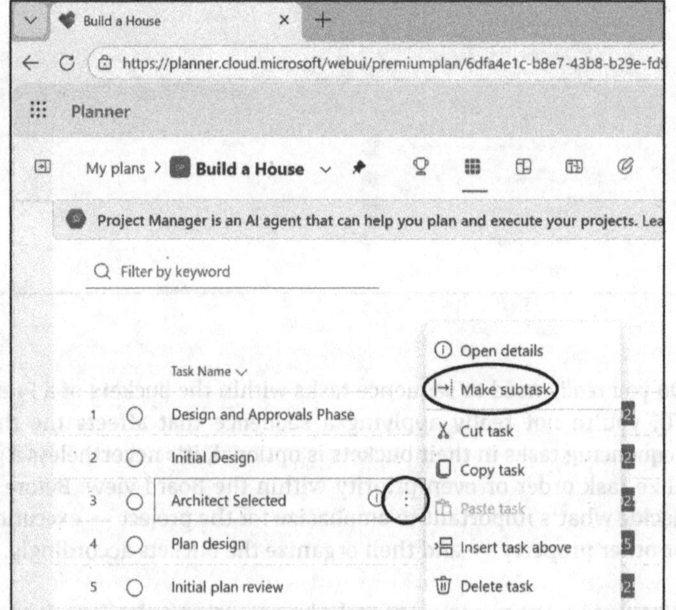

FIGURE 10-2: Use the Make Subtask command in a task's context menu to demote it to a subtask.

3. **Click the ellipsis to the right of the first task under the parent task and choose Make Subtask (see Figure 10-2).**

4. **Repeat Step 3 until you've demoted each of the subtasks as needed.**

5. **To move a subtask to a higher level, click the ellipsis beside the task name and choose Promote Subtask.**

As Figure 10-3 illustrates, the subtasks are indented under the parent task in Grid view, and you can collapse or expand the task and subtasks as needed. The parent task is also highlighted in bold.

When you demote the first subtask within a parent task, it can take a few seconds to complete. If you don't see an immediate result, wait a few seconds for the change to appear. Planner is doing some things in the background to create the nesting structure.

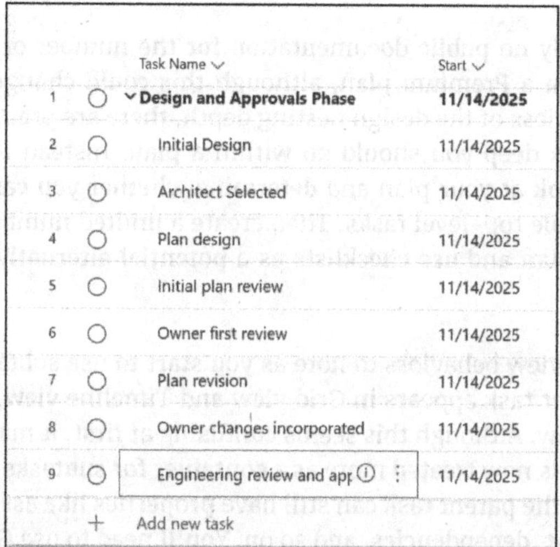

FIGURE 10-3:
The Design Phase task now includes six subtasks.

TIP

You can still use checklists in Premium plans, giving you an additional layer of granularity for both tasks and subtasks, or an alternative to using subtasks for simple plans.

Note you can create multiple levels of subtasks under a parent task. Figure 10-4 shows the Build a House plan with three task levels. The first parent is titled Design and Approvals Phase. Initial Design is a subtask of Design and Approvals Phase. The Initial Design task then includes four of its own tasks.

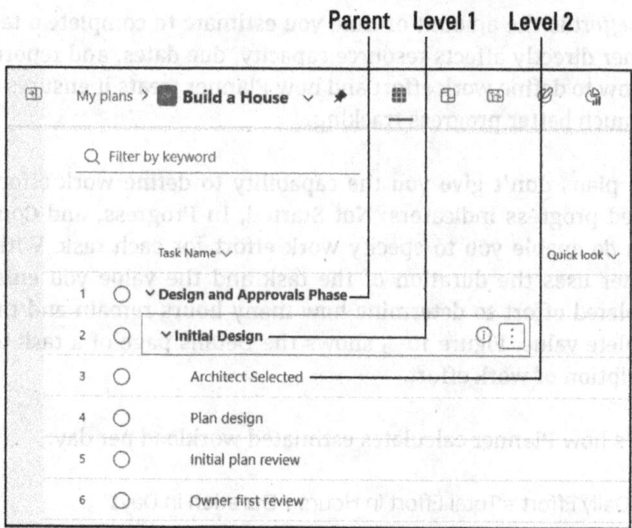

FIGURE 10-4:
In a Premium plan, you can create nested subtask levels.

There is currently no public documentation for the number of nested levels you can create within a Premium plan, although this could change by the time you read this. Regardless of the design nesting depth, there are practical limitations to how many levels deep you should go within a plan. Instead of creating several nested levels, look at your plan and determine whether you can break up parent tasks into multiple top-level tasks. Then create a limited number of nested levels to support the plan and use checklists as a potential alternative to deeper nesting levels.

There are some view behaviors to note as you start to use subtasks. For example, although a parent task appears in Grid view and Timeline view, it doesn't appear in the Bucket view. Although this seems confusing at first, it makes sense because that parent task is now treated more as a container for subtasks rather than a task itself. However, the parent task can still have properties like assignment, priority, labels, a checklist, dependencies, and so on. You'll need to use Grid view or Timeline view to view and edit the details of a top-level parent task.

When you need to add a new task, whether it's intended to be a parent or a subtask, you can click the ellipsis beside the task name and choose Insert Task Above. This inserts a blank task above the selected one. You can then drag the task into a different location in the sequence if needed, or demote/promote it as needed.

Defining Work Effort

Work effort is the amount of time you estimate to complete a task. Work effort in Planner directly affects resource capacity, due dates, and reporting. Understanding how to define work effort and how Planner treats it ensures accurate planning and much better progress tracking.

Basic plans don't give you the capability to define work effort. They also have limited progress indicators: Not Started, In Progress, and Completed. Premium plans *do* enable you to specify work effort for each task. With Premium plans, Planner uses the duration of the task and the value you enter in the task for completed effort to determine how many hours remain and the task's percent-complete value. Figure 10-5 shows the Details page of a task to help clarify this description of work effort.

Here's how Planner calculates estimated workload per day:

Daily Effort = Total Effort in Hours ÷ Duration in Days

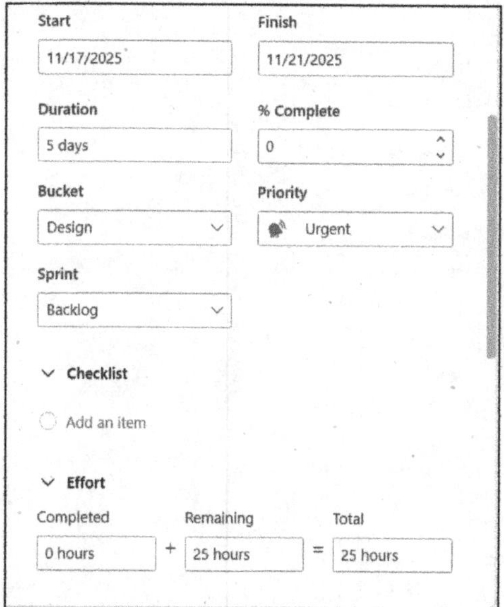

FIGURE 10-5:
Use a task's
Details page to
enter work effort
and progress.

Assume a task spans five days like the one shown in Figure 10-5. The duration is derived from the start and finish dates. Under Effort, you enter **25 hours** in the Total field. Initially, the task shows 0 percent complete. Planner determines the daily effort in the Duration field as five days based on the following calculation:

$$\text{Daily Effort} = 25 \div 5$$

Next, under Effort, you enter **5 hours** in the Completed field. After a slight delay, Planner calculates the remaining hours as 20 and updates the % Complete field to 20 percent. The result appears in Figure 10-6.

Here are some tips for defining and working with task effort in Planner Premium:

PREMIUM

>> Always start by setting the start and finish dates for a task.

>> Verify the Duration value that Planner calculates, as well as the Total hours.

>> If possible, avoid changing the start and finish dates, but if you do change them, verify that the Duration and Total hours change appropriately.

>> When entering progress, update the Completed field under Effort. Don't change the % Complete field.

>> Visualize effort and duration across tasks in Timeline view.

>> Create dependencies where appropriate to sequence tasks without creating overlapping workloads.

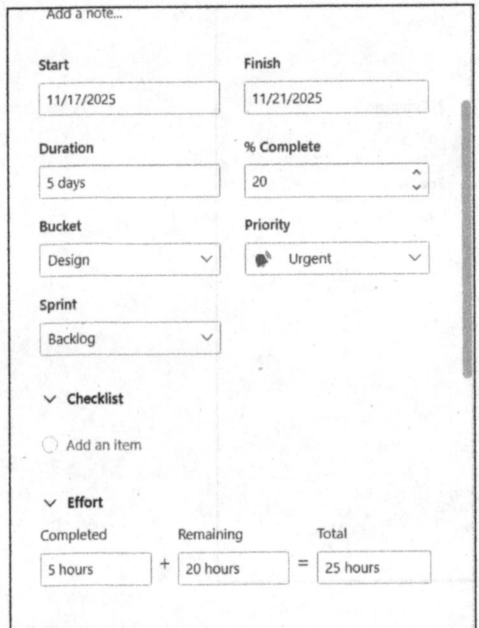

FIGURE 10-6: Planner calculates % Complete based on duration and completed effort.

You can override the % Complete value by entering a percentage directly. However, you should rely on Planner's ability to calculate percent complete whenever possible.

TIP

Managing Task Dependencies

PREMIUM

The term *task dependency* refers to the relationship between tasks where the start or completion of a task depends on the start or completion of another task. Task dependencies enable you to build flow requirements into a plan, defining when tasks must start or finish and how they relate to other tasks in the plan. For example, if you're building a house, you need to finish the foundation slab before you start framing the walls. If the house has a basement or crawl space, you need to finish installing the entire floor system on the foundation before you start erecting the wall system. These are task dependencies, and task dependencies become a necessity for task planning as project complexity grows.

The following sections explain the different types of dependencies and how you create and manage them in Planner.

Understanding types of dependencies

There are different types of dependencies that define the relationship between tasks:

>> **Finish-to-Start:** Task B starts when task A finishes. This is the default dependency.

>> **Finish-to-Finish:** Task B finishes when task A finishes.

>> **Start-to-Start:** Task B starts when task A starts.

>> **Start-to-Finish:** Task B finishes when task A starts.

>> **Lead/Lag:** This dependency type defines a lead time or lag time between tasks. *Lead time* schedules task B to start before task A by a certain span of time. *Lag time* schedules task B to start a certain span of time *after* task A finishes.

In Planner Plan 1, you can only create Finish-to-Start dependencies. Other types of dependencies require Planner and Project Plan 3 or Planner and Project Plan 5. Planner Basic does not directly support dependencies.

Task dependency is a Premium feature requiring Planner Plan 1 or higher. Even so, Planner Basic's flexibility enables you to track dependencies using some manual methods. The following section explains how.

Finding workarounds for Basic plans

Planner Basic does not support any task dependencies. You can, however, use some creative workarounds to identify dependencies in a Basic plan. As with progress and priority for a task, you can use labels, checklists, and comments to build a process within the plan for defining dependency. Here are some examples:

>> **Checklists:** Create a set of checklists for a task that identifies the dependent tasks. Using the construction analogy I use at the beginning of this section, the First Floor System task would have a checklist item called Foundation, because the foundation must be completed before the crew starts framing the floor.

>> **Comments:** Add a comment when you complete a task indicating that depending tasks can now start. Others who are responsible for those tasks will receive a comment notification and know that they can now start their tasks.

>> **Labels:** Add labels to a task that identify its dependent tasks. For example, create a label like, "Waiting on task A."

These methods don't give you the ability to view task dependencies in a cohesive way like you can with Premium plans, but they can still be useful for identifying dependencies when you have Planner Basic.

Creating and updating dependencies

When you're working with a Premium plan, you have the capability to create task dependencies. Figure 10-7 shows a plan called Build a House, with several top-level tasks for different phases of the project, starting with the design and working through the other construction phases. The Initial Design task contains several subtasks that are dependent on others.

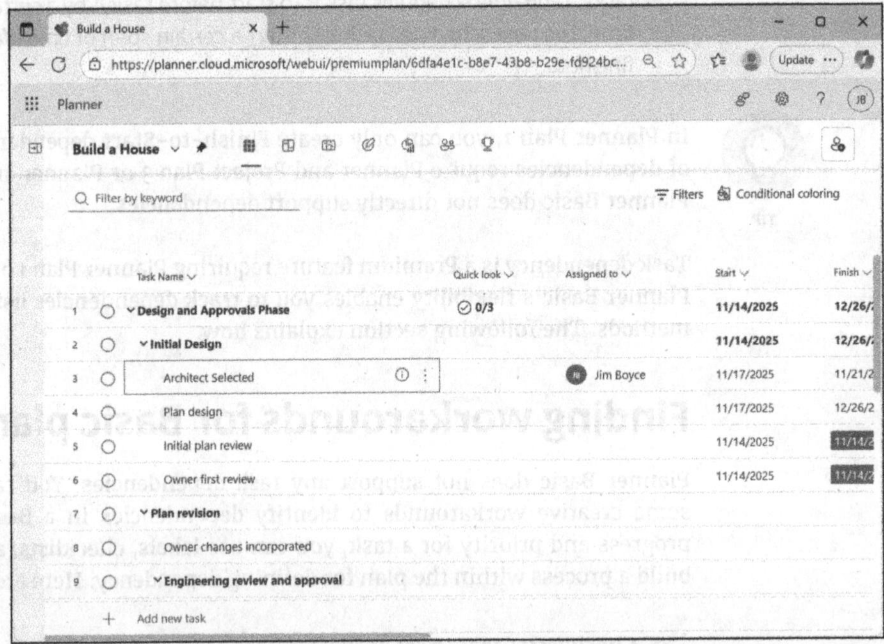

FIGURE 10-7:
The Build a House plan is full of dependent tasks.

The following list indicates the order in which the tasks need to be completed (a very simplified version):

1. **Architect selected:** The owner or general contractor chooses an architect for the project.

2. **Plan design:** The architect creates the initial design.

3. **Initial plan review:** The architect reviews the plans with the buyer.

4. **Plan revision:** The architect revises the plans as needed based on buyer feedback.

5. **Engineering review and approval:** The structural engineer, HVAC engineer, and other design professionals review and approve plans.

6. **Owner sign-off:** The buyer approves the plan.

7. **Construction drawings:** The architect provides construction drawings for the general contractor.

Every project is different, and there are potentially several other steps that take place in the design phase, but this list gives us enough to go by to start creating some dependencies. The following steps assume you have the Build a House plan I describe earlier with the tasks created in the Design bucket:

1. Open the "Plan design" task, which is the second task in the Initial Design phase.

2. On the task's Details page, click Add Dependency (see Figure 10-8).

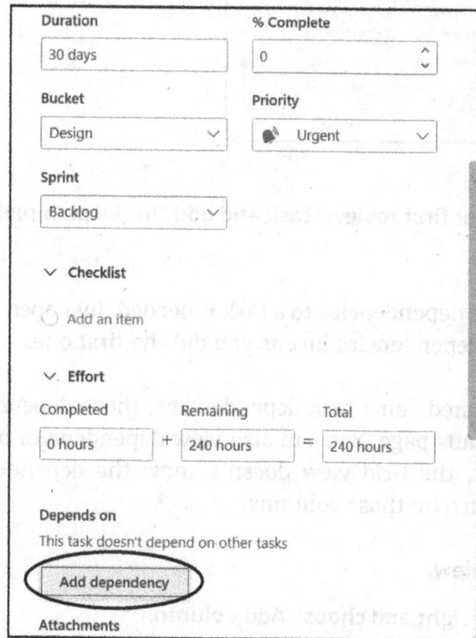

FIGURE 10-8: Click Add Dependency to search for a task.

3. Scroll through the resulting task list, choose "Architect selected," and then close the Details page.

4. **Open the "Initial plan review" task.**

5. **Click Add Dependency and choose Plan Design (see Figure 10-9).**

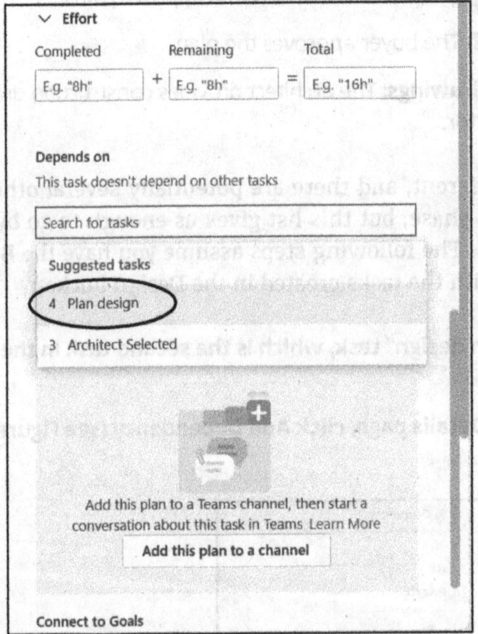

FIGURE 10-9:
Choose the
dependent task.

6. **Open the "Owner first review" task and add "Initial plan review" as the dependency.**

You can add multiple dependencies to a task if needed. Just open the task's Details page and add other dependencies just as you did the first one.

TIP

Now that you've created some task dependencies, those dependencies will show up on the task's Details page. You can also view dependencies in other ways. For example, by default, the Grid view doesn't show the dependent or depending tasks, but you can turn on those columns:

1. **Open the Grid view.**

2. **Scroll to the far right and choose Add Column.**

3. **Choose Depends on and Dependents.**

Those columns show the task number and task type, as shown in Figure 10-10 (I hid several columns for clarity). Unfortunately, that isn't very useful in the sense that you likely won't know what the dependent and depending tasks are based solely on their task numbers. A better way to view dependencies is to use Timeline view, shown in Figure 10-11.

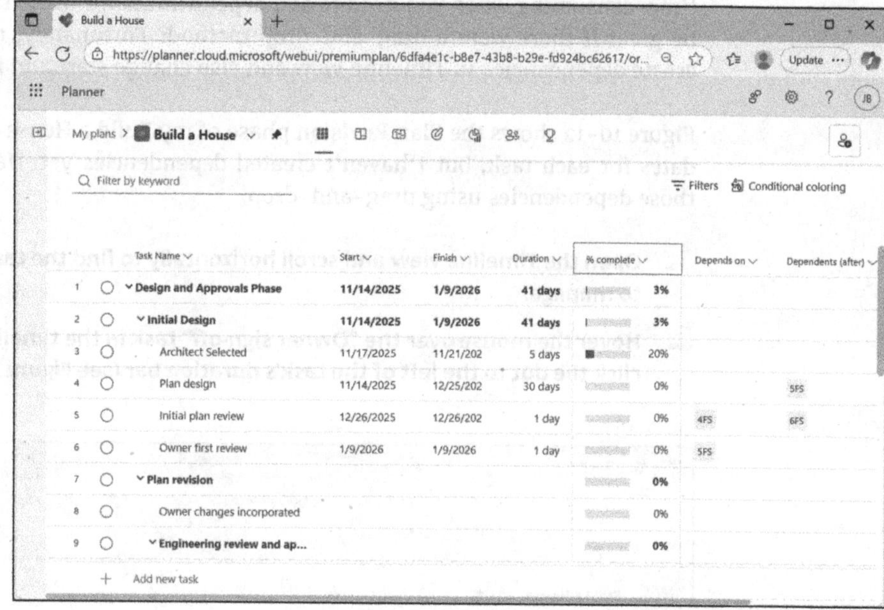

FIGURE 10-10: Dependency columns in Grid view.

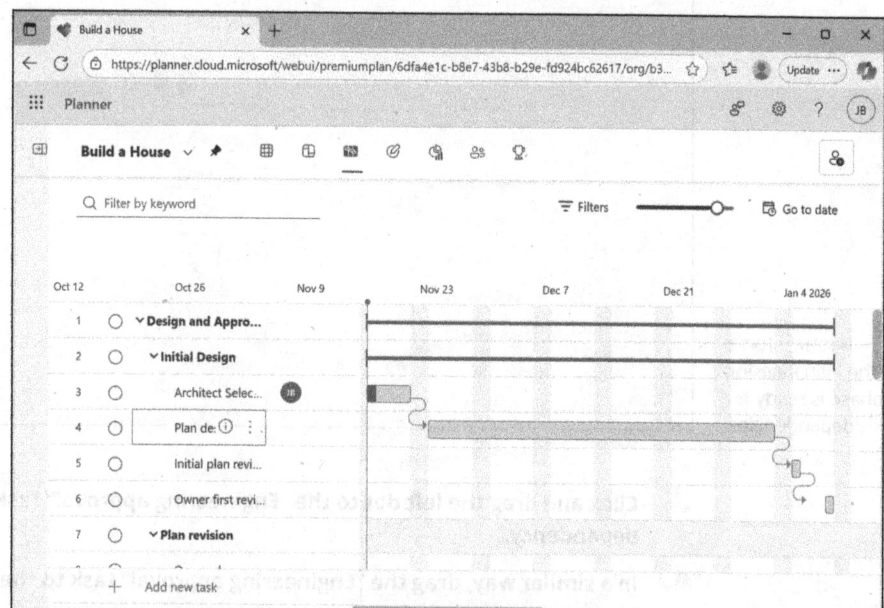

FIGURE 10-11: Timeline view shows task dependency.

Managing dates and dependencies in Timeline view

Adding dependencies through each task's Details page isn't difficult, but it can be time-consuming when you need to add dependencies for many tasks. Wouldn't it be great if there were a drag-and-drop method? Fortunately, there is. You can create dependencies in Timeline view and also change start and finish dates.

Figure 10-12 shows the Plan Revision phase of my Build a House plan. I've set the dates for each task, but I haven't created dependencies yet. Here's how to add those dependencies using drag-and-drop:

1. Open the Timeline view and scroll horizontally to find the tasks you want to manage.

2. Hover the mouse over the "Owner sign-off" task in the timeline pane and click the dot to the left of the task's duration bar (see Figure 10-12).

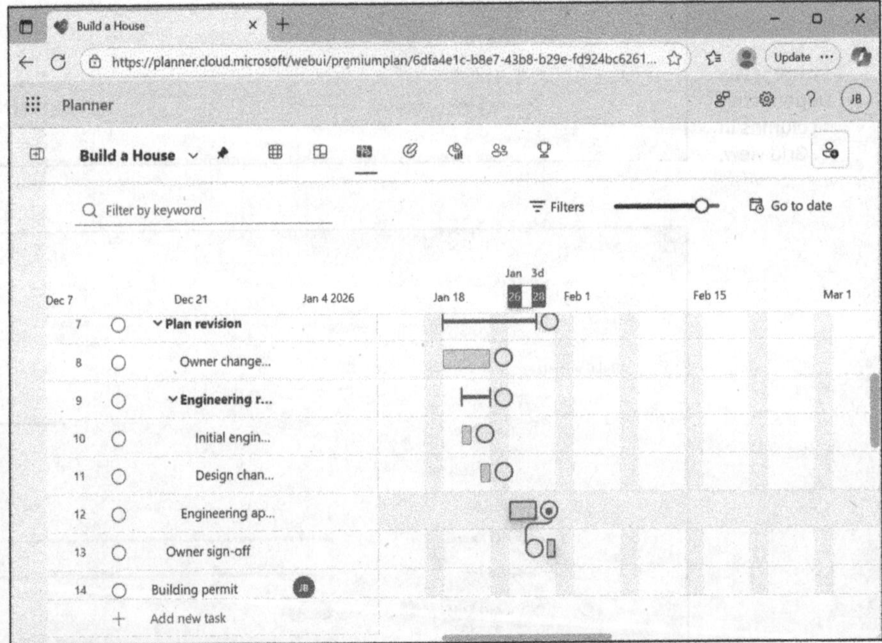

FIGURE 10-12: The Plan Revision phase is ready for dependencies.

3. Click and drag the left dot to the "Engineering approval" task to create a dependency.

4. In a similar way, drag the "Engineering approval" task to the "Design change complete" task.

5. **Drag the "Design change complete" task to the "Initial engineering review" task.**

Figure 10-13 shows the resulting dependencies.

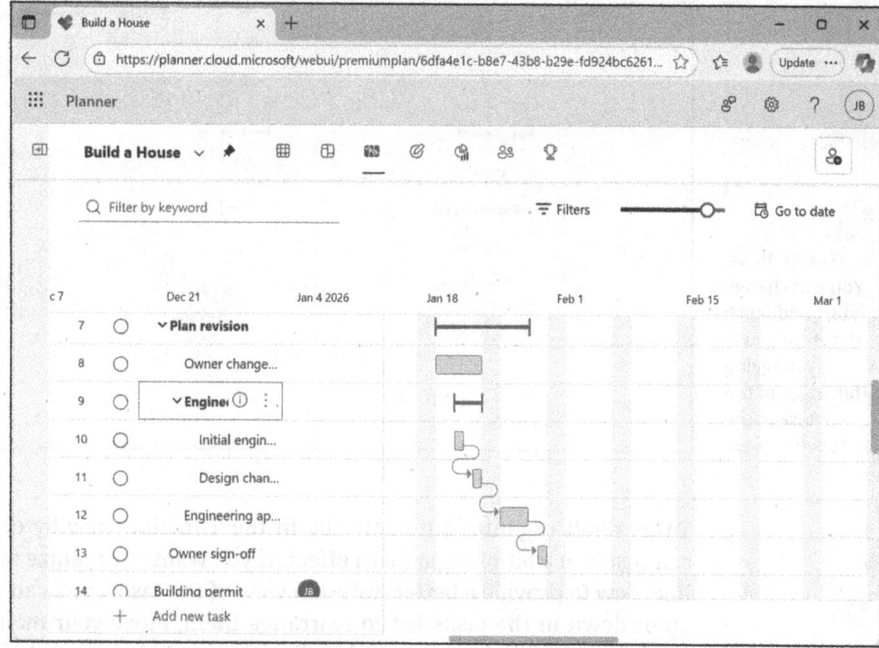

FIGURE 10-13: Several dependencies have been added using drag-and-drop.

In addition to adding dependencies in Timeline view, you can also change finish dates for tasks. For example, the Design change complete task shown in Figure 10-13 is too short — it'll take a couple of weeks to incorporate some engineering changes based on the structural engineer's feedback. So, I need to move the finish date out by a week. I could open the task's Details page and make the change there, but instead I'll change it directly in Timeline view:

1. **Open Timeline view and scroll to find the "Design change complete" task.**

2. **Hover the pointer over the right edge of the task's timeline bar and then click and drag the edge of the task to the right until the finish date reads "Jan 30."**

3. **Release the pointer.**

You can change the start and finish dates for a task using this same method. To change the start date, drag the left edge of the task's timeline bar to the left until the displayed start date indicates the desired date (see Figure 10-14).

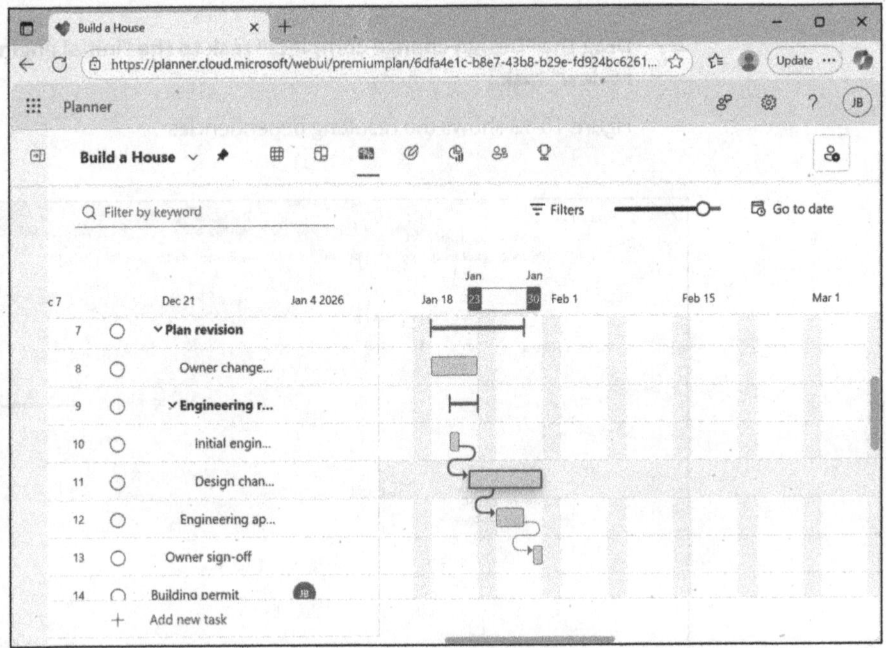

FIGURE 10-14:
You can change start and finish dates for a task by dragging the edges of the task in the Timeline view.

Tasks don't organize automatically in the Timeline view by dependency, which can result in a bit of a spaghetti effect. If you want to organize tasks on the Timeline view to provide a better waterfall view of the tasks, you can simply drag them up or down in the tasks list to rearrange them. Move your mouse to the far-left column to highlight the drag handle (six dots) and then drag the task to the desired location in the list.

Keep in mind that rearranging the order of tasks in this way renumbers the task. Drag task five above task four, for example, and they switch numbers (five becomes four and vice versa). Consider the potential impact to the plan based on how (if at all) you're using the task numbers before you start rearranging them.

Chapter **11**

Adding Structure with Templates

Setting up the structure of a simple plan doesn't take a lot of effort or time. Many simple plans are relatively generic in the sense that they either follow a very simple process or follow no prescribed process at all. If you're building a weight-loss plan, for example, you probably don't need a lot of structure to the plan — just clear goals and tasks and, of course, the willpower to follow through!

As plan complexity grows, the need for structure becomes paramount. Complex projects without a clearly defined structure and corresponding process will rapidly devolve into chaos — if they get off the ground at all. Starting out with a refined and thoughtful structure for a plan also saves an enormous amount of time required to establish goals, milestones, buckets, and tasks.

This chapter explores how to use templates in Microsoft Planner to quickly structure plans and explains how to tailor those plans to your own needs.

Exploring Use Cases for Templates

Templates enable you to quickly add structure to a new plan. They do this by providing a preconfigured structure based on plan and project type. Templates also help you focus your work through that same structure, as well as save a significant amount of time and increase efficiency in establishing the framework for a new plan. The following sections explore these concepts.

Quickly building structure

First and foremost, templates enable you to quickly build a plan structure based on the contents of the template. Building that structure is the key purpose for Planner templates and sets the foundation for the benefits that you derive from using templates. I like the house-building analogy I use in other chapters, so let's employ that here to illustrate the function that templates perform.

Every successful building project's plan has a structure to it. The plan's structure can vary somewhat depending on the type of building you're constructing, but it's generally structured something like this (excluding financial components):

>> **Design and approval:** The architect creates the design and gets approvals from structural engineers where appropriate and from the buyer.

>> **Contracting:** The general contractor is selected; they're responsible for subcontracting each trade (framing, drywall, and so on).

>> **Permitting:** This body of work involves plan reviews by a building official; documentation from the architect and structural engineer; possible revisions; and then issuance of the building permit, wiring permit, plumbing permit, and so on.

>> **Site work:** The site is prepped for the construction, including removal of existing structures, grading, excavation, and other preparation.

>> **Utilities:** Sewer, water, electricity, gas, and other utilities are roughed in to their respective locations.

>> **Foundation:** The foundation is completed, whether basement, crawl space, or slab on grade.

>> **Framing:** This body of work encompasses all structural framing (floors, walls, rafters, and so on).

>> **Fenestrations:** Doors and windows are installed.

>> **Roofing:** The roof structure is applied (shingles, roofing tiles, and so on).

>> **Electrical:** Interior and exterior wiring are completed.

>> **Plumbing:** Plumbing is completed.

>> **HVAC:** Installation of heating, ventilation, and air-conditioning systems is completed.

>> **Insulation:** All the insulation is installed in wall cavities, attic, and other areas.

>> **Exterior finishes:** Siding, paint, and other exterior finish work are done.

>> **Interior finishes:** Painting, trim, flooring, and fixtures are installed.

>> **Landscaping:** Finish grading, planting, and other exterior finish work are completed.

>> **Final inspections, tests, and issuance of a certificate of occupancy:** Prescribed testing is completed, final inspections are done, and a certificate of occupancy is issued by building officials.

Think of each bullet in the list as a bucket in your plan. Although that list *loosely* follows a timeline, there are no finish-to-start dependencies between the buckets. Often, multiple trades are at work on the project at the same time. For example, the plumbers and electricians may be roughing in their respective areas at the same time. Also, each bucket represents a large and detailed body of work of its own, with many dependencies between tasks across multiple buckets.

You can imagine that building a plan for a house project from scratch would be a Herculean effort. You would spend days just building the core structure for the plan. Here's where templates come to your rescue. Whether you're creating a plan to build a house or accomplish some other type of project, using a template that already has that structure built into it enables you to quickly set up a plan containing all the key elements needed for the project.

Saving time and increasing efficiency

The ability to quickly structure a plan using a template doesn't just mean that the plan starts out containing the main buckets, tasks, and other structure that you need for a complex project. It also can save a massive amount of time in setting up the plan. Consider the time you would put into building your first home construction plan. Can you imagine doing that from scratch each time you build a house? Whatever the project type, templates can remove a very significant amount of work from your planning process. That improves efficiency for the project, saving time, resources, and money that can be spent in other areas of the project.

Focusing your work

Another benefit of using templates is that they enable you to focus your work. They do this first by taking away the drudgery of building the structure for your plan, enabling you to focus on the details of the plan, not the structure.

Templates also bring consistency, which heightens focus in the key areas of the project by bringing predictability and clarity of responsibility to the project. This benefit grows as teams become accustomed to the same structure project after project.

Using the Included Templates

Planner includes several predefined plan templates in a mix of Basic and Premium plans. The following sections provide a very brief overview of the Basic and Premium templates that are included with Planner.

Note: The templates described in this chapter are the ones included with Planner as this book is in development. Other templates will likely be added over time.

Planner includes "empty" Basic and Premium options for creating a new plan. These options create plans with very limited structure and with features targeted for the specific plan type. They're useful when you want to experiment with the base features of Planner and to understand the differences in plan structure that the templates provide. They're also useful as starting points for building your own templates.

Basic templates

Planner offers five templates using Basic features, and these templates cover a small range of use cases. The Basic templates included in Planner are as follows:

>> **Simple Plan:** The Simple Plan template creates a plan structured primarily on task progress and focused on simplicity. It's useful for tracking personal day-to-day tasks, planning a personal event like a birthday party, organizing a small community event, or quickly capturing tasks for other projects.

>> **Project Management:** The Basic Project Management template provides a framework for creating plans that leverage common project-management processes for organizing, planning, and delivering projects. The template offers a simpler structure and greater simplicity than the Premium Project Management template (see the next section), making it useful for smaller projects where advanced features like dependencies aren't necessary. Consider using this template for community projects and events, small business initiatives, renovation projects, and event planning.

>> **Software Development:** The Software Development template is useful for smaller-scope application development or for projects focused on specific features or capabilities of a larger project. Some suggested used cases include tracking bugs, developing a specific feature, lightweight sprint planning, documenting projects, and testing and performing quality assurance.

>> **Business Plan:** The Basic Business Plan template is targeted at planning business opportunities and building a plan to execute on those opportunities. The template builds a structure that enables business owners, entrepreneurs, and aspiring impresarios to explore and define an opportunity, build a business approach to landing the opportunity, determine what success looks like, and develop a long-range plan to achieve the goals of the opportunity.

» **Employee Onboarding:** The Employee Onboarding template offers a structured approach to onboarding new employees to an organization or team, capturing the key tasks that need to be completed. Although the template is targeted to onboarding new employees to an organization, you could also customize it for similar activities like adding students to a class or adding an existing employee to a new team.

Premium templates

Planner includes several templates for creating Premium plans. Many of these templates build on the Basic templates described in the preceding section, leverage Premium features, and offer additional structure for more complex project planning. For example, the Premium Project Management template adds additional buckets, additional tasks and subtasks, and defined durations to enable you to start from a detailed plan.

The Premium templates included with Planner include the following:

» **Project Management:** The Premium Project Management template builds on the same structure as the Basic version, and adds a significant number of tasks and subtasks, defined task durations, and dependencies. The result is a detailed framework for fleshing out your project plan. The preestablished durations and dependencies that it provides streamline and speed up project planning.

» **Commercial Construction:** The Commercial Construction template provides a framework for crafting a construction plan modeled after established industry standards and best practices. It provides several work buckets preloaded with relevant tasks, each with target durations and dependencies where appropriate. The template is a great starting point for many different types of construction and can be tailored to specific types like commercial buildings, residential buildings, multifamily buildings, public buildings, and other project types.

» **Software Development:** The Premium Software Development template provides a preconfigured set of buckets to facilitate building a plan for a software development project. These buckets define the scope of the project; capture and analyze requirements; incorporate design and development phases; provide for testing and training; capture documentation; and plan for pilot, deployment, and post-implementation phases.

» **Sprint Planning:** A *sprint* is an iteration of a defined set of tasks or deliverables for a project that happens within a specific period of time (sometimes referred to as *time-boxed*). A sprint typically encompasses a fixed period anywhere from one to four weeks in length. The Sprint Planning template leverages agile development principles to enable you to break down complex projects into small, manageable bodies of work; balance workload across teams and resources; and adapt each sprint based on frequent reviews and feedback.

>> **Sprint Retrospective:** The Sprint Retrospective template goes hand in hand with the Sprint Planning template. The Sprint Planning template models the sprint itself, and the Sprint Retrospective template models the feedback and analysis for the sprint. In a nutshell, a sprint retrospective captures what went well, what did not, opportunities for improvement, and action items to capture in the next sprint.

>> **Project Retrospective:** Just as a sprint retrospective provides a review of a project sprint, a project retrospective provides a review of a project. The Project Retrospective template enables you to quickly build a consistent and repeatable framework for project closeout. The plan provides the framework for capturing highlights and challenges, recognizing teams and individuals for their accomplishments during the project, lessons learned, and takeaways for subsequent projects.

>> **Business Plan:** The Premium Business Plan template builds on the Basic Business Plan template. It includes the same buckets as the Basic template and adds the buckets Identifying Materials and Supplies, Re-evaluate Risks and Rewards, and Finalize Business Opportunity. The template includes additional tasks and subtasks, estimated task durations, and dependencies, all distributed across multiple phases.

>> **Goals and Objectives:** The Goals feature in Planner enables you to define, track, and align measurable objectives to desired strategic outcomes. Where tasks describe discrete actions aligned to outcomes for specific phases of a project, goals track to much broader organizational objectives. Tasks drive toward achievement of goals, and Planner enables you to set, measure, and visualize achievement against those goals.

>> **Employee Onboarding:** The Premium version of the Employee Onboarding template uses the same buckets as the Basic version but includes additional tasks in each bucket for a more detailed onboarding experience. It adds task duration and dependencies, and the additional tasks are tailored for organizations with special requirements like background checks and employee badging.

>> **Training Plan:** The Training Plan template is geared toward the orientation and training activities related to onboarding a new employee or team member. It provides a great starting point for building your own orientation and training plan for new employees.

>> **Simple Project:** The Simple Project template is just what its name implies — a very simple plan that you can (and will need to) modify to structure simple projects. The template contains just two buckets called Summary #1 and Summary #2, each with a few generic tasks. At first blush, the template may not seem very useful, but it's handy for experimenting with a Premium plan that already has some task durations and dependencies created up front.

>> **Marketing Campaign:** The Marketing Campaign template provides a full framework for planning a marketing campaign, taking the project from developing the campaign concepts, through all the planning phases, to campaign release and retrospective.

>> **Event Planning:** The Event Planning template offers a detailed, preconfigured plan, complete with tasks, suggested durations, and dependencies already built in for planning and hosting a major event. It takes the project through all phases for venue selection, event programming, promotions, catering, post-event closeout, and all other aspects of planning and execution.

>> **CRM Pipeline:** You can use the CRM Pipeline template to build and manage a customer relationship list with buckets for the various phases of a customer or client outreach. (*CRM* stands for *customer relationship management.*) For example, assume yours is a small organization and you want to build a list of potential customers and track them through a sales motion for a new service or product. You can use the CRM Pipeline template to store their contact information and move each contact through the various stages from identifying leads to completing the sales motion.

>> **Help Desk Tickets:** The simple Help Desk Tickets template treats tasks as request records. For example, if someone contacts you because they can't open their email, you create a task for that issue, assign it to someone to investigate, and then they move it through the buckets, which include Incoming, Investigating, In Progress, Testing, On Hold, and Done.

Customizing Templates

Using the templates that are available with Planner is a great way to save time and build consistency in your project plans. But they aren't just for one-off projects. Templates enable you to define a plan structure that you can use repeatedly for types of projects that recur over time or that follow a common process established by your organization. This section describes the benefits of using templates and existing project plans to build a library of your own templates.

Creating your own templates

Imagine that you've just come off a six-month project where you built and used a Planner plan to drive the project to completion. Now, leadership tells you that a similar project is coming down the pipeline for you to manage. Do you really want to re-create what you used over the past six months, or would you rather lather, rinse, and repeat? You'd choose the latter, of course!

Planner currently doesn't give you the means to save a plan as a template, but you can copy a plan to create a new one. However, there's one problem: All the tasks and other information in the plan are also copied to the new one. You'd likely have a lot of work to do to turn that new project back to the starting point.

Instead, create customized but *empty* plans and use them like templates to create preconfigured and fully ready plans that don't have any project-specific data in them yet. You can then modify them as needed to suit the upcoming project without having to re-create the plan from scratch.

Here's a general process to follow:

1. **Determine whether you need Premium features for the plan.**

2. **Determine if there is an existing Planner template that fits the majority of your needs.**

3. **Start a new plan using the selected template.**

4. **Add buckets, checklists, labels, custom fields, and other plan elements as needed.**

5. **Give the plan a name that indicates it's a customized, self-created template.**

   This way, you and others will know that it's a baseline template rather than an active project plan.

6. **If you have a backup solution available, back up the finished personal template; otherwise, ensure that you're the only one who has access to it.**

7. **When it's time to use your personal template for a new project, create a copy of the plan, rename the copy accordingly, and use it for your next project.**

The following sections describe how to customize your personal template.

Adding custom fields

After you've chosen a template or created a starting point for your new plan, you'll likely want to customize it. For example, you'll want to add custom task fields for specific use cases. Using the Help Desk Tickets template described earlier in this chapter, you may want to add fields for Case Number, Severity, Support Topic, and so on.

Follow these steps to create a custom task field:

1. **Open the plan to be customized.**

2. **Open the Grid view.**

3. **Scroll to the far right and click Add Column.**

4. **If none of the existing fields fits your needs, click New Field.**

   The New Field dialog box (shown in Figure 11-1) appears.

New field

This field will be available to all tasks in the plan.

Type

Text

Field name

Name your field

Create   Cancel

5. **From the Type drop-down list, select a field type (Text, Date, and so on).**

   Choose a field type that fits the situation. When you need to force yourself or others to enter information consistently, choose the Choice field type. As Figure 11-2 illustrates, you can create a list of options from which the user can choose, enforcing consistency for quality of data and reporting. Examples include product names or types, technology type, application, and any other information that should be selected from a predefined list.

6. **In the Field Name box, type a name for the new field.**

7. **Click Create.**

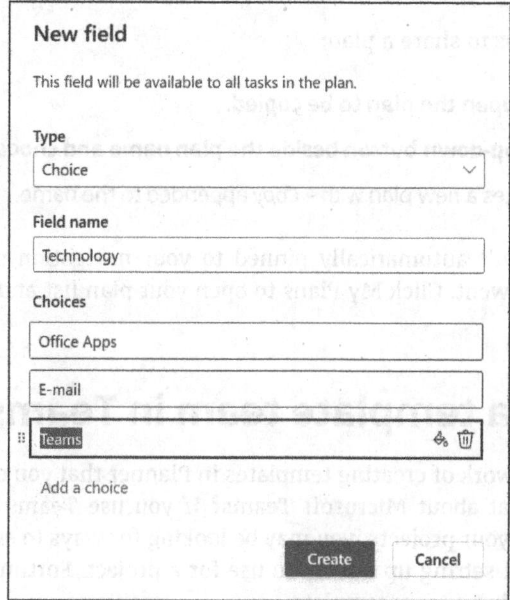

FIGURE 11-2:
Use the Choice field type to constrain data entry and enforce consistency.

REMEMBER

I can't stress too strongly how important it is to build consistency into your templates, but particularly those where you need to capture specific information that you want to track later. For example, if you're using the Help Desk Tickets template, it's very important that you create support topics like Email, Networking, Operating System, Hardware, VPN (short for *virtual private network*), and so on, not only to ensure that all cases are consistent in how they track the issue topic, but also so you can report on that data later. For example, you'll want to know what percentage of your cases are for networking, email, or other specific topics so you can identify commonalities and work to reduce user confusion or other common issues in each support area. If you don't force support staff to choose from an existing list, your data will rapidly become garbage. You can always add to the pick list when something new comes up.

Copying the plan

Copying a plan is easy. When you do so, Planner creates a plan with the same name and - *Copy* appended to the plan name. The new plan doesn't have any group assignment, so it's available only to you. This way you can create your own library of templates and share them when you need to start a new project, either for yourself or to be shared with others.

Follow these steps to share a plan:

1. **Locate and open the plan to be copied.**

2. **Click the drop-down button beside the plan name and choose Copy Plan.**

   Planner creates a new plan with – *Copy* appended to the name.

The new plan isn't automatically pinned to your navigation pane so you may wonder where it went. Click My Plans to open your plan list and locate the newly copied plan.

Creating a template team in Teams

You've done the work of creating templates in Planner that you can reuse for your projects, but what about Microsoft Teams? If you use Teams as a platform to manage and run your projects, you may be looking for ways to simplify and speed up the process of setting up a team to use for a project. Fortunately, you can do just that by creating a team template.

There are two ways to create a team template in Teams. If you're a Teams administrator, you can create a template through the Teams admin center. To create a template using this method, see Chapter 16. The following explains how to create a team from an existing template that was created by your admin team:

1. **In Teams, click the ellipsis at the top of the navigation pane and choose Your Teams and Channels.**

   Teams displays a list of all your teams.

2. **In the upper-right corner of the window, click Create Team.**

   The Create a Team dialog box appears.

3. **Click the More Create Team Options link.**

4. **In the navigation pane, click From Template.**

   Teams display a list of available templates (see Figure 11-3).

5. **Click the team to be used as the template for your new project team.**

6. **Click Use This Template.**

   The Create a Team dialog box appears.

7. **Enter the team name, description, team type and sensitivity, and click Create.**

8. **Open your teams and channels, click the ellipsis beside the newly created team, and choose Manage Team.**

9. **Make any final changes as needed and add members to the team.**

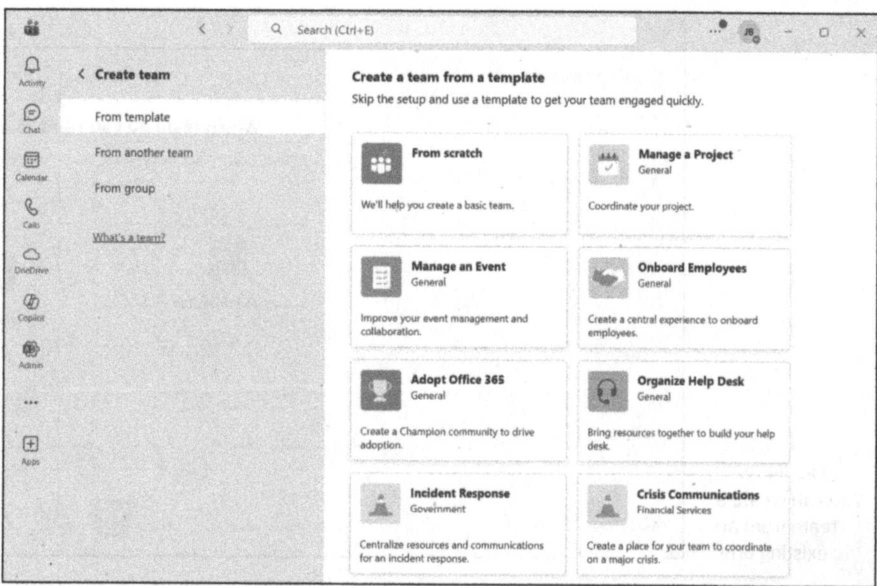

FIGURE 11-3: You can create a team from a template.

There is no feature in Teams for end users to create and save a team as a template, but that's also true for Planner. Instead of creating a Planner template, you create a preconfigured plan and save it for use later. The same concept applies for Teams. If you can't get an admin to create one for you, you create the team in Teams, add tabs and other features to set it up the way you want, and then create a new team using that one as your model.

The following assumes you've already created the "template" team, but if you haven't, you can use the same process to create it first:

1. **In Teams, click the ellipsis at the top of the navigation pane and choose Your Teams and Channels.**

   Teams displays a list of all your teams.

2. **In the upper-right corner of the window, click Create Team.**

   The Create a Team dialog box appears.

3. **Click the More Create Team Options link.**

4. **In the navigation pane, click From Another Team.**

   Teams display a list of your teams (see Figure 11-4).

5. **Click the team to be used as the template for your new project team.**

   A dialog box appears in which you set the team name and choose other options (see Figure 11-5).

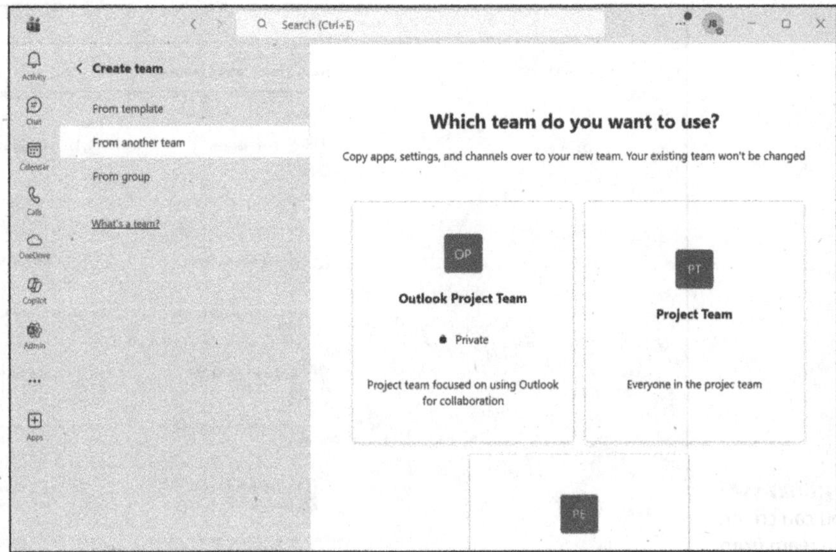

FIGURE 11-4:
You can create a team from an existing one.

FIGURE 11-5:
Planner appends [copy] to the end of the group name, but you can change the name.

6. **Enter the desired options, including choosing which items to include from the existing team in the new one.**

   You can also choose to include the members of the previous team in the new one.

7. **Click Create.**

Chapter **12**

Managing Goals, Sprints, and Milestones

Goals are a feature in Microsoft Planner Premium versions that enable you to define high-level goals and align tasks to make progress toward those goals. Although the Goals feature is not available in Basic plans, there are some workarounds you can implement in Planner Basic to track goals.

A *sprint* is a fixed period of time, usually from one to four weeks, during which a team plans, develops, and delivers a discrete set of tasks. Breaking up a longer project into sprints enables the team to better focus their efforts and achieve results with a smaller body of work. This chapter explains how to use the Sprints feature in Planner Premium.

Milestones are another feature included in Planner Premium versions but not in Planner Basic. Milestones are project checkpoints that serve as symbolic markers for key points in a project. Milestones can be linked to dependencies and, therefore, become part of the project's critical path.

This chapter explores the Goals feature and milestones in Planner Premium and offers workarounds for simulating both features in Basic plans.

Setting and Working with Goals

Everyone has personal goals. Get fit. Save money for college. Find a new job. Retire early. Many people don't put those goals on paper or fully consider what they need to do to achieve them, but the goals still float around in the backs of their minds and may drive decisions and actions that move them closer to achieving them.

Organizations have goals, too. These goals may be to cut costs in the fiscal year, increase revenue, diversify income streams, improve employee morale, and improve customer satisfaction, to name a few. Whether you're working on your own personal goals or tasked with managing goals for an organization, Planner's Goals feature is a great option for defining and tracking progress toward achieving those goals.

In Planner Premium, a *goal* is a discrete object type. Other Planner objects include buckets and tasks. In some ways, Planner goals may seem similar to buckets in that they can have tasks associated with them. But goals are not buckets. Instead, they provide a high-level object to which you can assign tasks. The difference between the two, therefore, is that goals represent strategic outcomes that are driven by tasks, while buckets are a tool for organizing tasks. This distinction becomes more apparent as you begin to work with goals, so let's dive in.

The following sections explore goals in Planner and how to create and manage them.

Creating and managing goals

Instead of building a goal-oriented plan from scratch for this discussion, let's use a template:

1. **Open Planner and at the bottom of the navigation pane, click New Plan.**

   The Create New dialog box appears.

2. **Scroll down and click See All Templates.**

   The Templates dialog box appears.

3. **Scroll through the list and click the Goals and Objectives template (see Figure 12-1).**

4. **Click Use Template.**

5. **Give the plan a name, pin it to your plans, and click Create.**

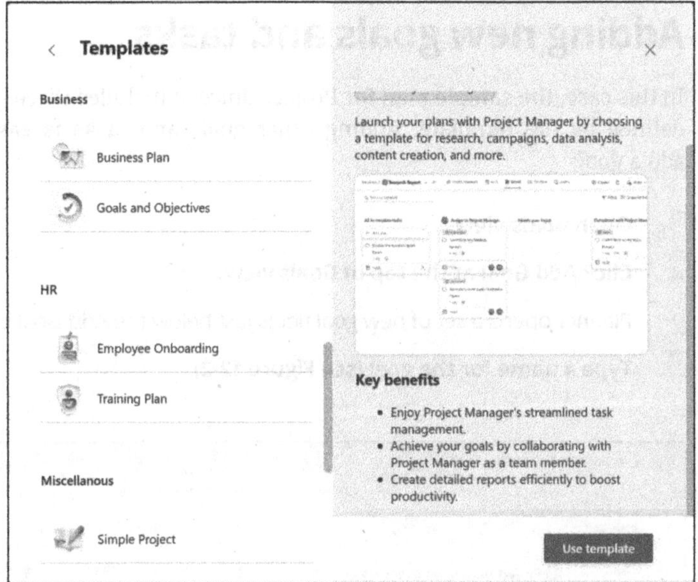

FIGURE 12-1:
Create a goal-oriented plan with the Goals and Objectives template.

For this example, I created a plan called Project Unicorn Goals and Objectives. The plan lays out the goals for the launch of a product called Unicorn, named after a fondly remembered internal Microsoft tool of the same name. The plan has three goals:

>> **Keep the product launch on budget and on time:** This goal includes several tasks aligned to product development and launch.

>> **Expand into international market:** Tasks aligned to this goal include exploring international privacy and compliance regulations, researching competitors, developing a marketing strategy, and other high-level activities that must be achieved to meet the goal.

>> **Launch version 2.0 within six months to improve user engagement and satisfaction:** This goal includes several tasks aligned to taking feedback from the initial release, reanalyzing the market, and planning the version 2.0 launch of the product.

Figure 12-1 shows the Goals view, which groups the tasks under their corresponding goals. You can work directly in this view with tasks (for example, marking them as complete by clicking the circle beside the task name). To open the task details, click the Open Details button to the right of the task name, or click the More Options button and choose Open Details. With the Details page open, you can manage the task's properties just like you do for tasks that are not associated with goals, like setting start and finish times, assigning the task, assigning it to a bucket or sprint, and so on.

Adding new goals and tasks

In this case, the sample plan for Project Unicorn includes three goals that are pre-defined by the template. Adding other goals and tasks is easy. Here's how to add a goal:

1. **Open Goals view.**

2. **Click Add Goal at the top of Goals view.**

   Planner opens a set of new goal fields just below the Add Goal button.

3. **Type a name for the goal (see Figure 12-2).**

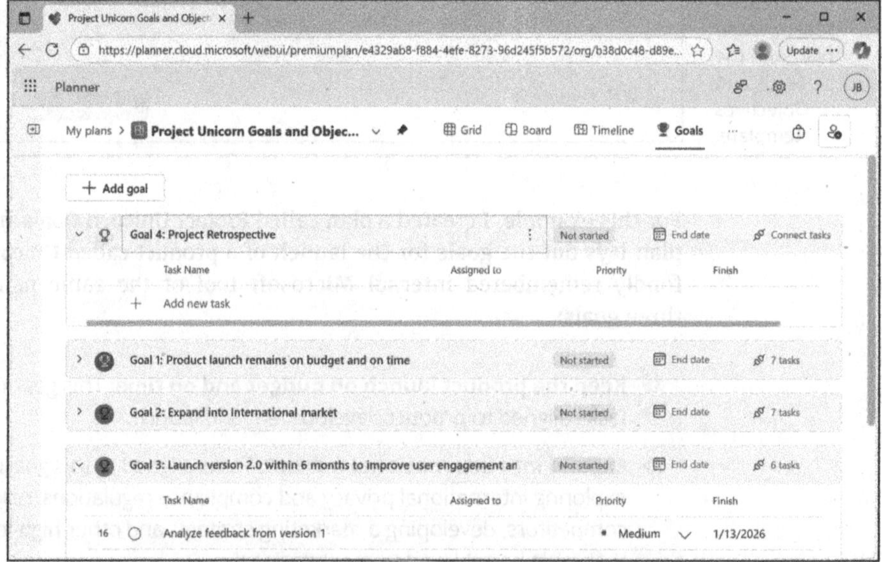

FIGURE 12-2:
Click Add Goal and type a name to create a new goal.

4. **Set other fields as needed and click outside the Add Goal area to close the goal properties and save the goal.**

Each goal should have tasks, so the next action is to add a couple of tasks for the new goal:

1. **Still in the goals plan, switch to Grid view.**

   You can also create tasks in Board view.

2. **Click Add New Task.**

3. **Enter the values for the task's properties.**

4. **Create other tasks for the new goal as needed.**

5. **Scroll to the bottom of Grid view, locate one of the new tasks, and click the More Options button to the right of the task.**

6. **Choose Connect to Goal, and then choose the goal to which the task will be linked.**

   Figure 12-3 shows an example.

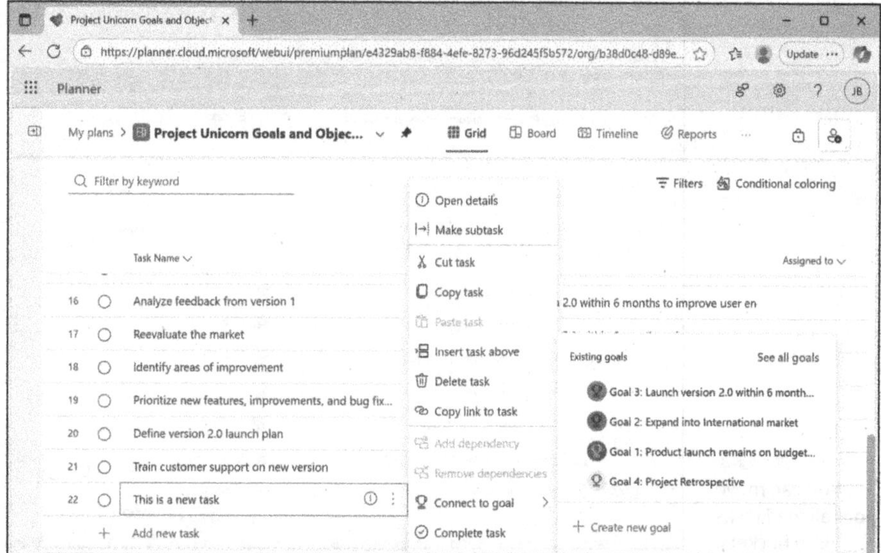

FIGURE 12-3: You can link tasks to a goal.

7. **Repeat the process for other tasks.**

Tasks in a goals plan are no different from tasks in other plans. You can work with them in the same way you work with tasks in other plans.

Going a step further

At this point, I have four goals, each with several tasks. I could stop at this point and just work with the tasks as they are, setting dates, adding priorities, adding labels, assigning the tasks, and so on. But I want to add some additional organizational structure to the plan. I do that by adding buckets and moving tasks into those buckets. I assume in the following steps that you know how to create buckets (if you don't, turn to Chapter 7). In this example, I've renamed Bucket 1 to

Project Management and created three more buckets named Development, User Acceptance Testing, and Launch:

1. **With the additional buckets created, open Grid view.**

2. **Scroll to find the task you want to move to a bucket.**

3. **Open the task's Details pane (see Figure 12-4).**

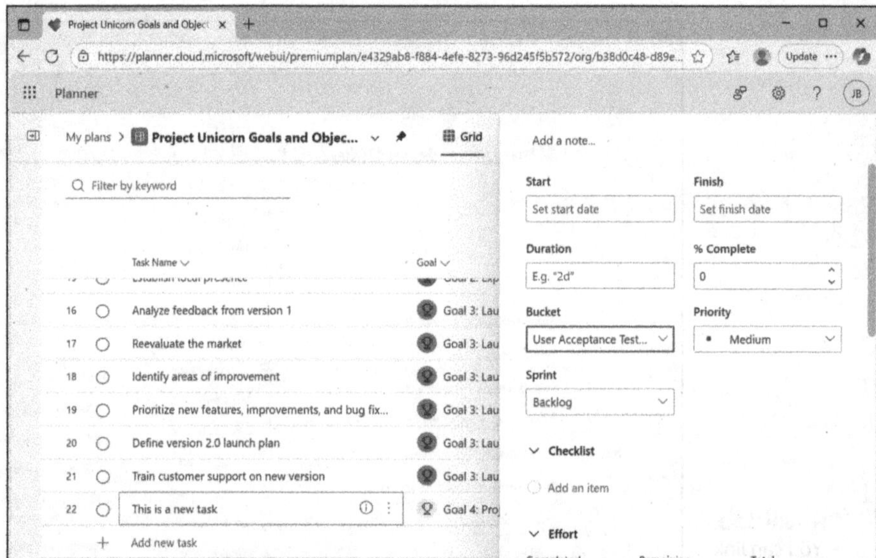

FIGURE 12-4:
You can move goal-aligned tasks into buckets.

4. **Click the Bucket drop-down list and choose the destination bucket.**

5. **Repeat the process for other tasks as needed.**

   Figure 12-5 shows an example of a task placed in the User Acceptance Testing bucket.

The previous steps reinforce the concept that buckets and goals are very different objects. Tasks can be grouped into different buckets even if they align to the same goal. This enables you to not only align tasks with the appropriate goals, but also leverage buckets to organize the tasks in ways that make sense for the project at hand.

With your goals and tasks created for your project, you can begin to track toward achievement of those goals. As tasks are marked as completed, they begin to contribute to the goal's progress. You can use Charts view in Planner to view task assignment, status, and progress. For more details on reporting in Planner, turn to Chapter 15.

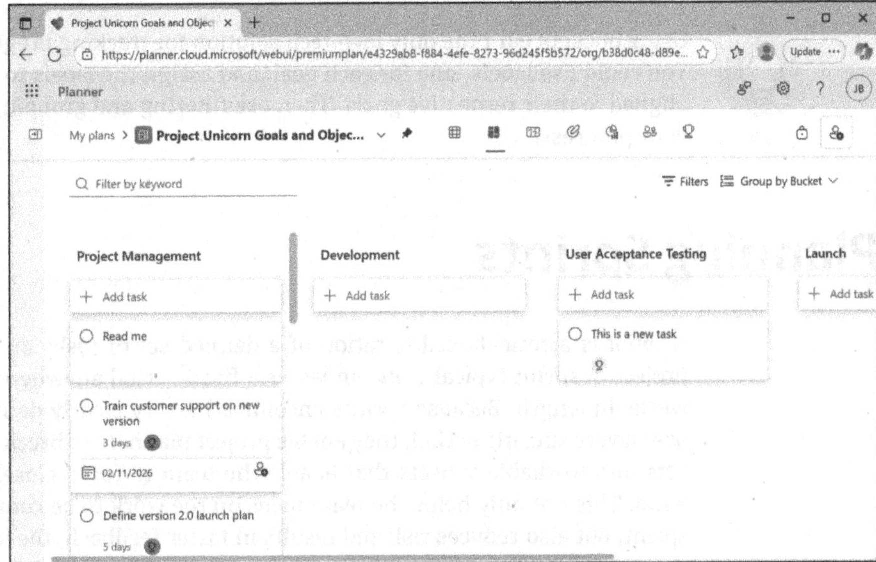

FIGURE 12-5:
A new task is placed in the User Acceptance Testing bucket.

Workaround for Planner Basic

The Goals feature, as mentioned earlier in this chapter, is a Premium feature requiring at least Planner Plan 1. Goals and the capability to link tasks to goals are not available in Planner Basic. So, what do you do if you want to use a Basic plan to track your goals? You use buckets!

This is what IT geeks like me call a *low-tech solution* because it's a very basic solution cobbled together without using any advanced features. It's only a little better than duct tape and chewing gum. Even so, it's a logical approach. Here's how it works:

1. **Create a bucket for each goal.**

2. **Create tasks for each goal and "align" them to goals by moving them into the appropriate goal bucket.**

3. **Track progress toward each goal based on the completion of the tasks within each bucket.**

It's certainly not an elegant solution, but you get what you pay for. Even so, this method can be very handy for tracking personal goals or goals related to small projects where a Premium plan isn't possible.

Using buckets isn't the only low-tech solution for tracking progress toward goals. You could use labels, one for each goal, and assign the labels to the tasks that are aligned to their respective goals. Then use filtering and grouping by label to track your progress.

Planning Sprints

A *sprint* is a time-boxed iteration of a defined set of tasks or deliverables for a project. A sprint typically encompasses a fixed period anywhere from one to four weeks in length. Because sprints encompass a very clearly defined body of work over a very specific period, they enable project planners to break up complex projects into workable subsets that enable the team to focus closely on the work at hand. This not only helps the team focus on the work to be completed during the sprint, but also reduces risk and results in faster feedback, the latter enabling the team to improve operationally from sprint to sprint.

Sprints follow a plan–execute–analyze–improve model, and include the following elements:

>> **Goal:** The outcome to be achieved during the sprint

>> **Tasks:** The work to be performed during the sprint to achieve the goal

>> **Fixed time box:** A defined timeline for the sprint (such as two weeks)

>> **Review:** A team review of what was completed

>> **Retrospective:** A team retrospective to identify successes and challenges with an outcome of how to improve the next sprint

Sprints are a Premium feature that requires Planner Plan 1 or higher. The following section explores how to create and use sprints in Planner Plan 1.

Planning sprints in Planner Plan 1

Planner Premium plans enable you to create and track projects based on the sprint model. Although the Sprints feature in Planner will likely change over time through additional development, sprints are a somewhat hidden feature in Planner because there is currently no dedicated Sprints view (contrary to what a lot of the online documentation says). I hope Microsoft will fix this soon, but fortunately, sprints are not buried too deep in Planner.

To begin experimenting with sprints, you first need to create a Premium plan. In this section, I use a plan created with the Sprint Planning template. As Figure 12-6 illustrates, the resulting plan includes four buckets:

>> **Project:** The tasks that the team will use to define the project effort; includes tasks like drafting user stories and prioritizing those by value

>> **Design:** Tasks that encompass the design and testing activities during the sprint

>> **Development:** The actual development activities

>> **Meetings:** Tasks for the review and retrospective meetings

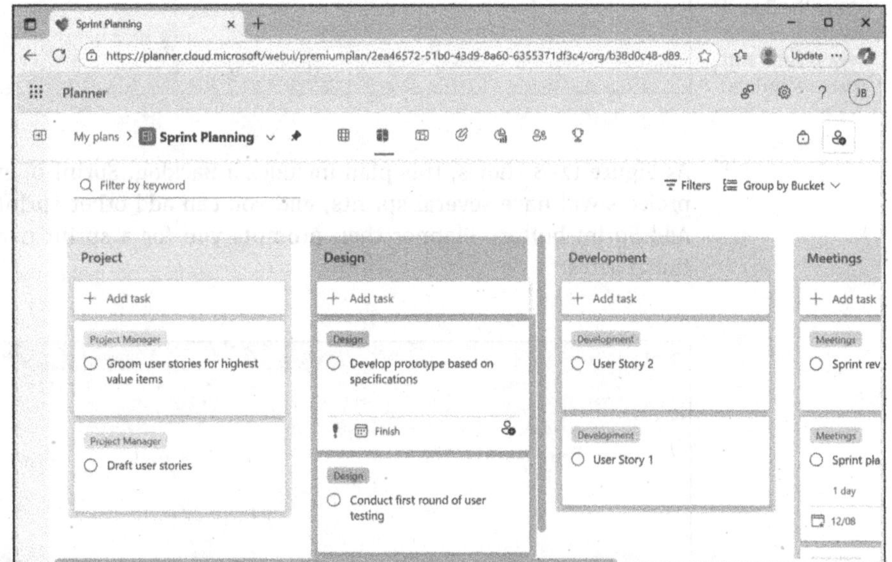

FIGURE 12-6: A plan created using the Sprint Planning template.

Now, let's go find those sprints:

1. **Open a Premium plan.**

2. **Open Board view.**

3. **At the right of the view, click Group by Bucket and choose Sprint (see Figure 12-7).**

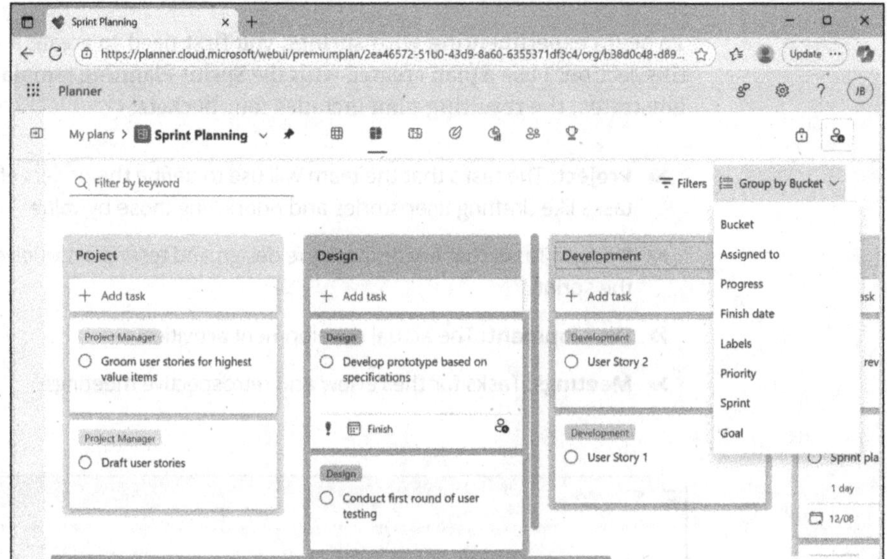

As Figure 12-8 shows, this plan includes a Backlog, Sprint 0, and Sprint 1. Most projects will have several sprints, and you can add other sprints by clicking the Add Sprint button. Planner then prompts you for a sprint name and start and finish dates.

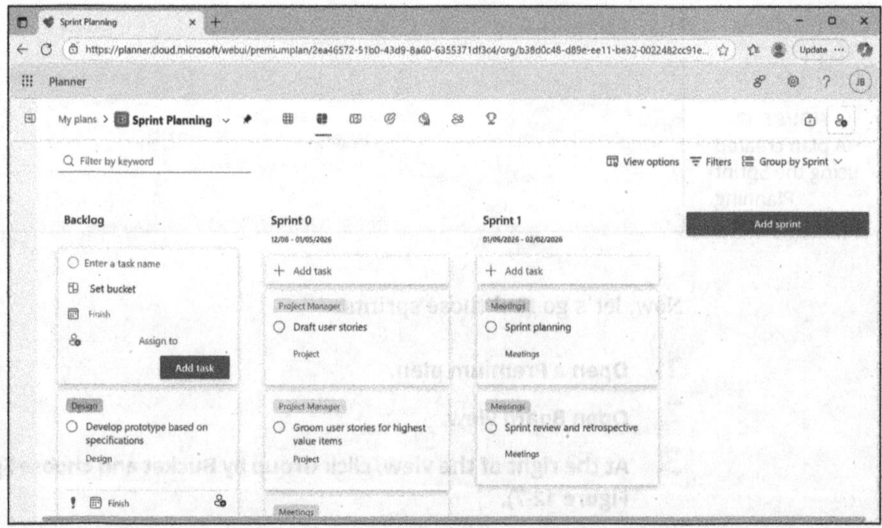

The following outlines the general steps for using sprints in Planner (the steps assume you're familiar enough working in Planner to accomplish these steps):

1. **Use a Premium plan.**

   First make sure you're using a Premium plan. Convert a Basic plan to Premium if needed, or just create a new Premium plan.

2. **Create a backlog bucket.**

   In the backlog bucket, build the backlog of tasks that need to be accomplished for the project. Flesh out the tasks with appropriate descriptions, checklists, dependencies, and other properties, and then prioritize them.

3. **Create the sprints.**

   Based on the backlog, create the various sprints required for the project.

4. **Plan the first sprint.**

   Using the prioritized list of tasks in the backlog, identify the tasks for the first sprint. Open the Details pane for each of these tasks and use the Sprint field to assign the tasks to the sprint.

5. **Run the sprint.**

   After aligning the appropriate tasks to the first sprint, work with the team to run the sprint (perform development and execute other tasks in the sprint).

6. **Review the sprint.**

   Host the sprint review meeting to go through all the sprint's tasks to review task completion status, review deliverables, and demo the outcome where appropriate.

7. **Perform a sprint retrospective.**

   Host the retrospective meeting to identify what went well, what didn't go well, and opportunities for improvement.

8. **Repeat.**

   Go back to Step 3 to build and run the next sprint.

How you run each sprint will vary somewhat based on the project, the team's working style, and other factors. Following is a set of recommendations for how to leverage Planner's features to run a sprint:

>> Update task status daily (Not Started, In Progress, or Completed).

>> Use checklists within the tasks to add structure and clarity and check off steps as they're completed.

» Add daily comments to help track status.

» As blockers come up, add them to the backlog.

» Use the task history to view the task status throughout the sprint.

» Routinely review status and adjust assignments where appropriate. Hold short, daily sprint meetings to keep the team advised and on track.

As the previous example illustrates, sprints are a separate object type from buckets. You can use buckets to organize tasks based on type, team, and so on. The sprints then define *when* the tasks will be completed. Leverage filtering and grouping in the various Planner views to switch between Board and Sprint views.

Planning sprints in Basic plans

This discussion of sprints in Planner would be incomplete without describing a workaround to enable you to model sprints in Basic plans. Maybe you'd like to model a small project using sprints, but you don't have a Premium license. Even personal projects can benefit from the sprint model, keeping you focused and on task.

If you've read the section, "Workaround for Planner Basic," earlier in this chapter, you know it covers how to model goals without a Premium license, so you can probably guess where this discussion is going. The low-tech solution is to use buckets!

Create a Basic plan for your project that contains a bucket called Backlog. Use this bucket to create and prioritize the tasks that will make up the project deliverables. Create a bucket for each sprint and name it appropriately (Sprint 0, Sprint 1, and so on). Move tasks from the Backlog bucket to each sprint bucket, and then work through the tasks in each sprint. At the end of each sprint, hold your own review and retrospective to determine next steps and improvement points.

Working with Milestones

Milestones in Planner are a Premium feature that enables you to create zero-duration tasks that represent milestones in your plan. Milestones aren't tasks per se but rather checkpoints within the project. They take their name from actual stone markers that have been used along roads for thousands of years to mark distance in miles.

The interstate system in the United States is a more recent example that includes mile-marker signs every mile to indicate the distance from the beginning of a segment to the current point. For example, when you cross a state line heading east, the first mile marker you encounter is mile marker 1. The marker numbers increase as you continue east, and they start back at 1 when you cross the next state line. The same is true for U.S. numbered highways and most state highways. Mile markers on north–south routes increase sequentially as you travel north.

Planner milestones represent significant points in a project's timeline, often marking the completion of a major project phase or deliverable or serve as a decision point. Milestones help you visualize these key points in Timeline view.

Milestones are another Planner Premium feature that will likely change over time. As I write this, there is no "Create Milestone" command or something similar. Instead, you create them indirectly from tasks (more on that shortly). Milestones are also not separate objects from tasks. Instead, they're tasks with zero duration, and they often have a dependency.

The following steps add a milestone to a plan I created with the Commercial Construction template, marking the completion of the first project phase:

1. **Open the plan in which you want to create the milestone.**

2. **Open Timeline view.**

3. **Click Add New Task at the bottom of the window.**

4. **Create a new task named Phase 1 Milestone.**

5. **Grab the task's grab handle and drag it into position as the last task in the first project phase.**

6. **Click the More Options button at the right of the task's name and choose Open Details to open the task's Details pane.**

7. **In the Details pane, choose a finish date for the milestone.**

   The duration will change automatically to reflect the specified finish date.

8. **Enter 0 in the Duration field.**

   This automatically sets the start date to the same date as the finish date.

9. **Optionally, click Add Dependency to add a dependency to the previous task in the sequence.**

10. **Close the Details pane.**

Figure 12-9 shows the results. The General Conditions phase now includes a task named Phase 1 Milestone, represented by a diamond on the timeline.

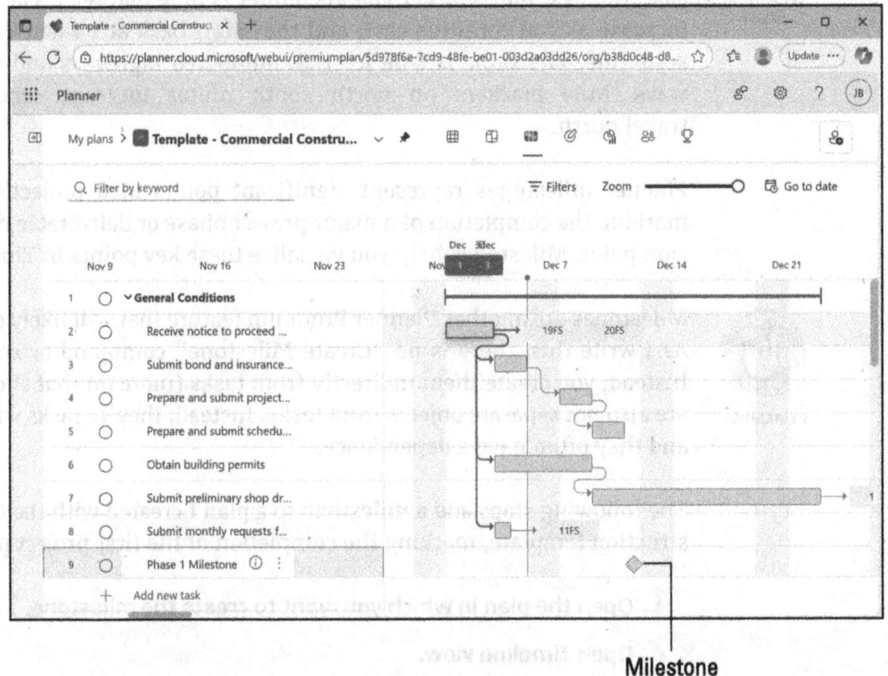

FIGURE 12-9: The Phase 1 Milestone is represented by a diamond in Timeline view.

Although adding a dependency for a milestone is optional, the milestone will be considered in the project's critical path only if it sits at the end of a dependency chain (for example, is dependent on the last task in a phase) or sits between two dependent tasks.

TIP

Although I have nothing to go on, my educated guess is that the Milestone feature in Planner will evolve to the point where milestones are discrete objects like Goals, with a means to create them directly.

4
Automating and Managing Microsoft Planner

Chapter **13**

Leveraging Automation and Integration

Microsoft Planner offers an easy-to-use interface and enables you to accomplish all project-planning tasks through its web app or through Microsoft Teams. But as part of the Microsoft 365 ecosystem, Planner integrates with several other Microsoft tools including Microsoft Outlook, Microsoft To Do, Microsoft SharePoint, and Microsoft Loop. It also can leverage Microsoft Power Automate for automating tasks. You can use Planner quite effectively by itself, but the capabilities offered by these other applications and tools can simplify your workday and empower collaboration with others.

This chapter explores these integration and collaboration features, beginning with how Power Automate and Planner can work together to automate and streamline tasks.

Automating Workflows with Power Automate

Planner includes a built-in connector that enables you to create automated interactions with Planner using Power Automate. You can perform actions like creating and organizing tasks, updating tasks, sending notifications, and more. These interactions can extend Planner's capabilities, as well as enable you to work more effectively and efficiently in Planner.

Introducing Power Automate

Power Automate is one component of Microsoft Power Platform, which comprises the following tools:

>> **Power Automate:** Create automated workflows across applications, services, and systems. You can use Power Automate to automate repetitive tasks, move data between systems, trigger actions when specific events occur, and extract data using artificial intelligence (AI).

>> **Microsoft Power BI:** Transform and model data from numerous sources to create interactive dashboards and reports that visualize the data in multiple ways. Gain insights into your data using these visualizations and AI-driven queries.

>> **Microsoft Power Apps:** Create low-code and no-code applications using drag-and-drop and other simple techniques. These apps can run on many different platforms.

Power Automate is a cloud–based workflow service that automates actions across common apps and services through the creation and execution of *flows.* You can create these flows using little or no code, making Power Automate a great tool for streamlining and automating tasks without having much, if any, programming experience. Flows can post messages, send notifications, create and manage files, automate email actions, query databases, submit forms, perform actions in online services like SharePoint and Microsoft OneDrive, automate actions in a desktop application, manage approval workflows, and much more.

Exploring Power Automate capabilities in Planner

Power Automate integrates with Planner to enable you to:

>> Create and update tasks.

>> Add checklist items.

>> Assign tasks to a user.

>> Move tasks to a bucket.

>> Mark tasks as complete.

>> Trigger flows when tasks are created or completed.

Here are several use cases for integrating Planner with Power Automate:

>> **Creating tasks from a form:** Perhaps users create a help-desk ticket by filling out an online form. When they submit the form, Power Automate can create a task in the help desk's project plan and, based on the contents of the form, assign the task to a bucket named Incoming Requests, assign a priority, and assign the task to someone.

>> **Creating tasks from a SharePoint list:** Maybe your organization uses a SharePoint list to enable users to request maintenance activities. When a new item is added to the list, Power Automate can create a task using the fields in the list to assign a task name, dates, target team, and other values.

>> **Automating repetitive plan creation activities:** Maybe your project team accepts project requests from an online source like a SharePoint list or form. Power Automate can create a new project plan, populate the plan with initial tasks and other properties, and assign the initial discovery task to the lead project manager.

>> **Sending notifications:** Planner sends notifications, by default, to a group when a new plan is created and to you when you're assigned a task or one of your assigned tasks is due soon or overdue. These are just some of the notifications that Planner sends without any additional automation. In situations where you need a custom notification, such as when tasks move between buckets or priority is changed, you can use Power Automate to generate the notification.

>> **Creating approval workflows:** Maybe you have a large project coming up and want to integrate reviews and approvals into the plan. You can create a flow in Power Automate that triggers an approval process when a task with certain properties is created. When the approval is provided, Power Automate marks the task as completed and notifies you of the approval.

>> **Syncing tasks with other systems:** In some situations, you may want to sync data between Planner and other systems. For example, perhaps you use Planner to manage team tasks, but you need to update Microsoft Dynamics 365 (a customer relationship management, or CRM, tool) when a task is assigned or completed. You can create a flow to automate that action.

>> **Cleaning up project plans:** If you're working with a relatively large project with lots of participants and tasks, keeping the plan organized can be a challenge. You can use Power Automate to automatically move tasks to a Completed bucket when team members mark them as completed. The flow can perform other cleanup tasks as well.

The next section walks through a hands-on example of creating a Power Automate flow.

Trying it out

The sample Power Automate flow described in this section creates a task in Planner when an item is added to a SharePoint list. You might use this solution when you use a SharePoint list to enable users to submit help-desk tickets, maintenance requests, or other types of requests. The example assumes you have access to create a simple SharePoint list and can create a plan to use in the example. Start by creating a plan for the support requests and assign it to a Microsoft 365 group. The plan should have the following properties:

» **Plan type:** Create a Basic plan.

» **Title:** Name the plan My Support Requests.

» **Buckets:** Create a bucket called Incoming Requests.

You could include other buckets, but for simplicity, these plan elements are the only ones that this example uses. Now comes the fun part. Let's start by creating the SharePoint list:

1. **Create a Microsoft 365 group named Project Team.**

   Creating the group creates a SharePoint site by the same name. You can instead use an existing group and site of which you're a member.

2. **Open the SharePoint site, click New, and choose List.**

   The How Would You Like to Start? page appears (see Figure 13-1).

3. **Click List.**

4. **Type the name Help Desk Requests (se Figure 13-2), and click Create.**

5. **Close the Next Steps pane, click the Settings gear icon, and choose List Settings.**

   The list's Settings page appears.

6. **Click Create Column and create a choice column named Priority.**

7. **In the Type Each Choice on a Separate Line field, enter three choices: High, Medium, and Low (see Figure 13-3).**

8. **Click Yes to require that the field contains information, which forces the users to choose an option.**

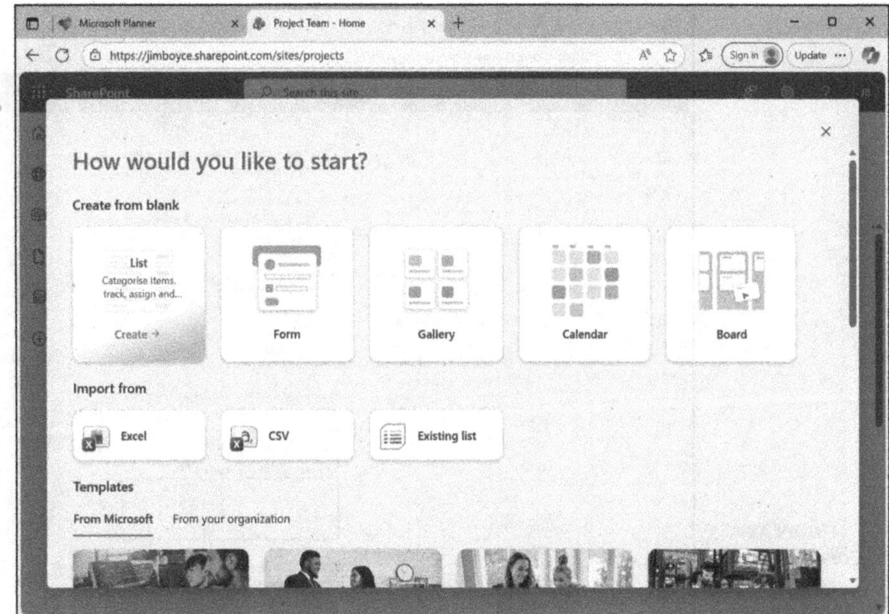

FIGURE 13-1:
The How Would You Like to Start? page.

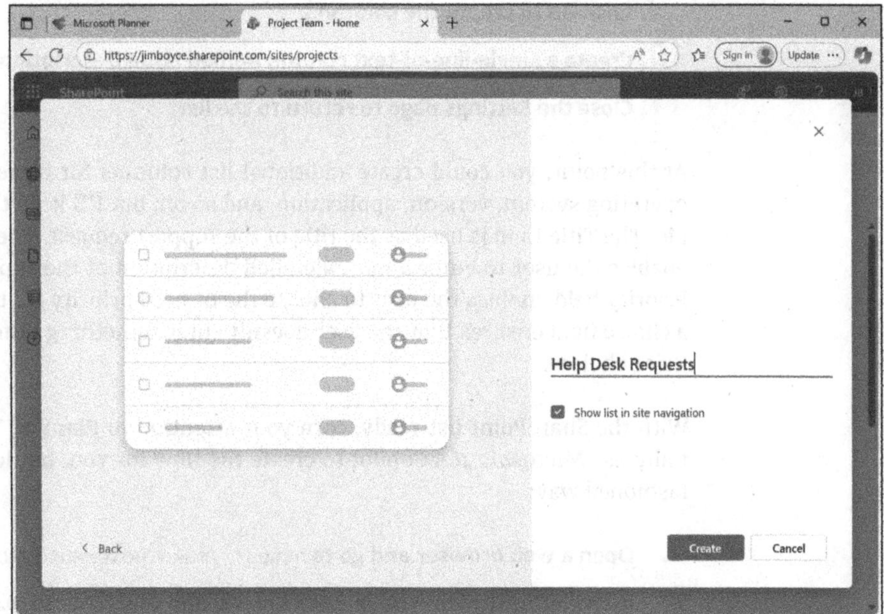

FIGURE 13-2:
Name the list Help Desk Requests.

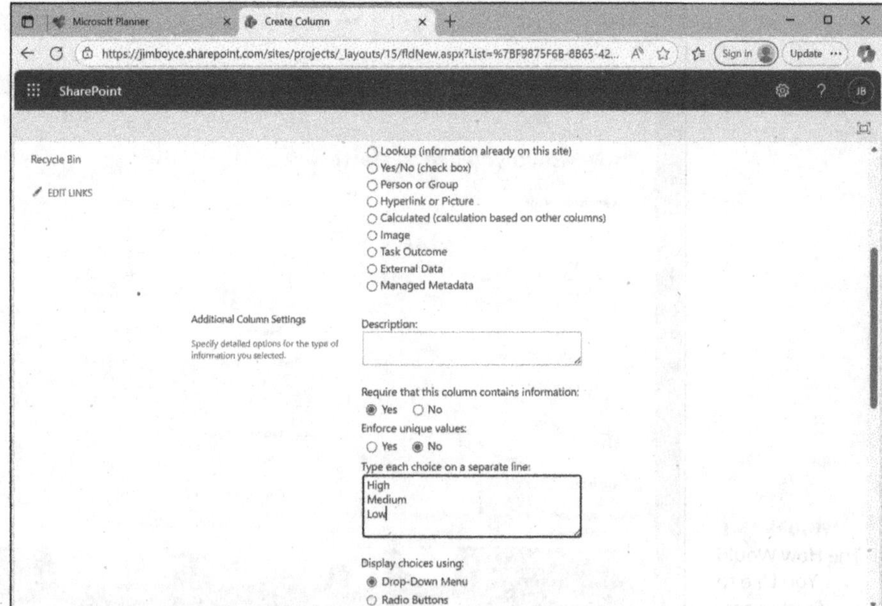

FIGURE 13-3:
Create a choice
column with
three choices.

9. **Click OK to create the column.**

10. **Create a single-line-of-text column named Issue Description.**

11. **Close the Settings page to return to the list.**

At this point, you could create additional list columns for more information like operating system, version, application, and so on, but I'll keep this example simple. The Title field is used as the title of the support request. The Description field enables the user to enter a more detailed description of the support request. The Priority field enables the user to choose the desired priority for the request. Using a choice field ensures that the user doesn't enter something other than one of the three choices.

With the SharePoint list ready, turn your attention to Planner. You could potentially use Microsoft 365 Copilot to create the flow for you, but let's do it the old-fashioned way:

1. **Open a web browser and go to** https://make.powerautomate.com.

2. **In the left navigation pane, click Create, and then choose Automated Cloud Flow (see Figure 13-4).**

   The Build an Automated Cloud Flow window appears.

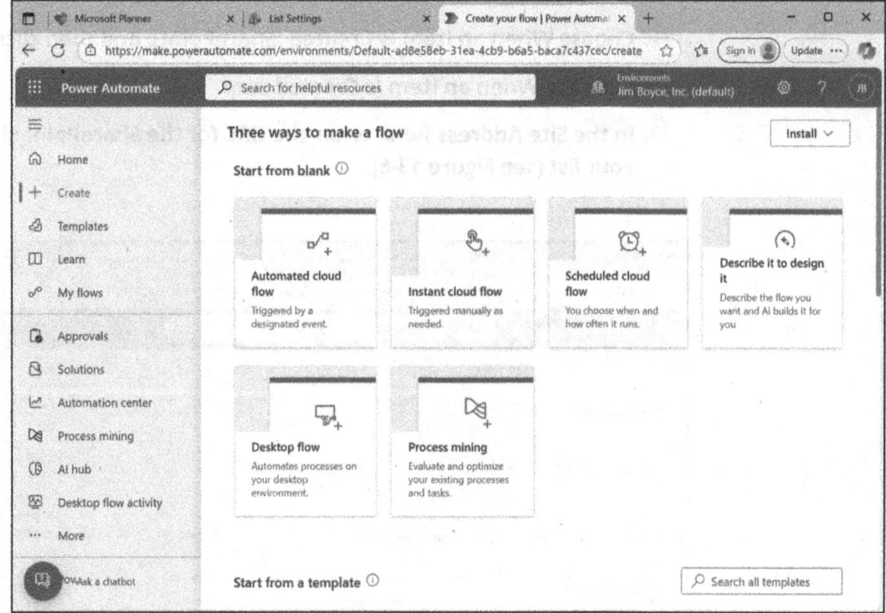

3. **In the Flow Name field, type** Support Request to Planner **(see Figure 13-5).**

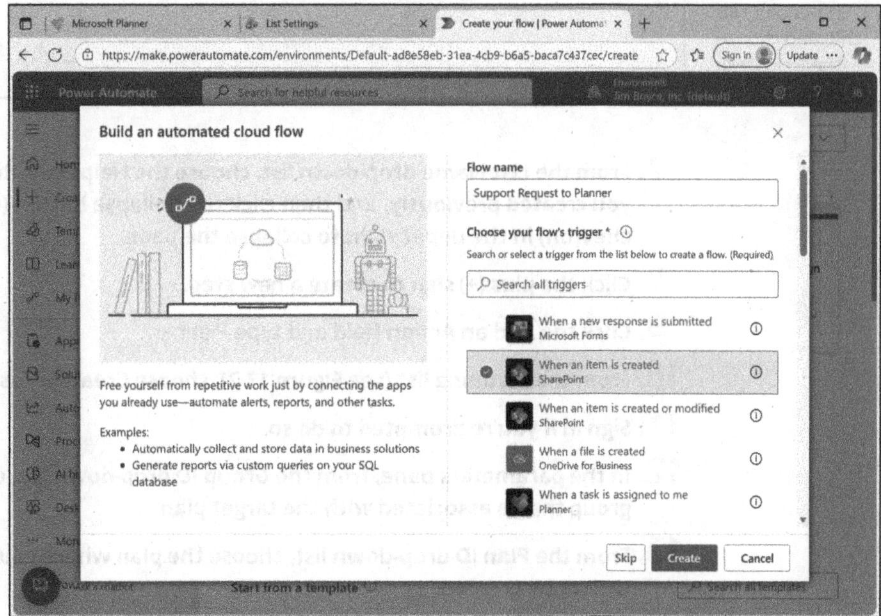

4. Choose **When an Item Is Created (SharePoint)**, and then click **Create**.

5. Click the **When an Item Is Created** step.

6. In the **Site Address** field, enter the URL for the SharePoint site that hosts your list (see Figure 13-6).

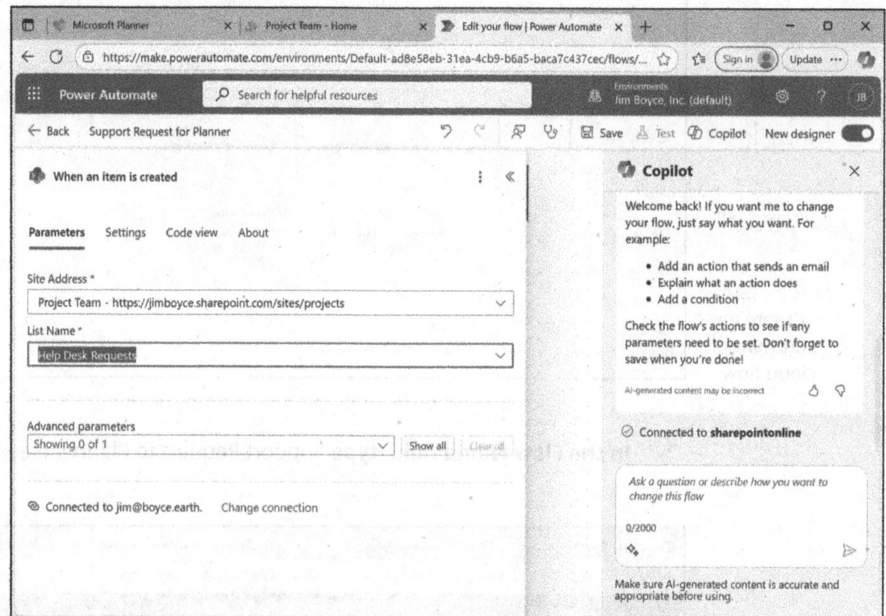

FIGURE 13-6:
Enter the SharePoint site URL and choose the Help Desk Requests list.

7. From the **List Name** drop-down list, choose the Help Desk Requests list you created previously, and then click the Collapse button (double-left chevron) in the upper right to collapse the pane.

8. Click the plus (+) sign to create a new step.

9. Click the **Add an Action** field and type Planner.

10. From the resulting list (see Figure 13-7), choose **Create a Task**.

11. Sign in if you're prompted to do so.

12. In the parameters pane, from the **Group ID** drop-down list, choose the group that is associated with the target plan.

13. From the **Plan ID** drop-down list, choose the plan where you want the task to be created.

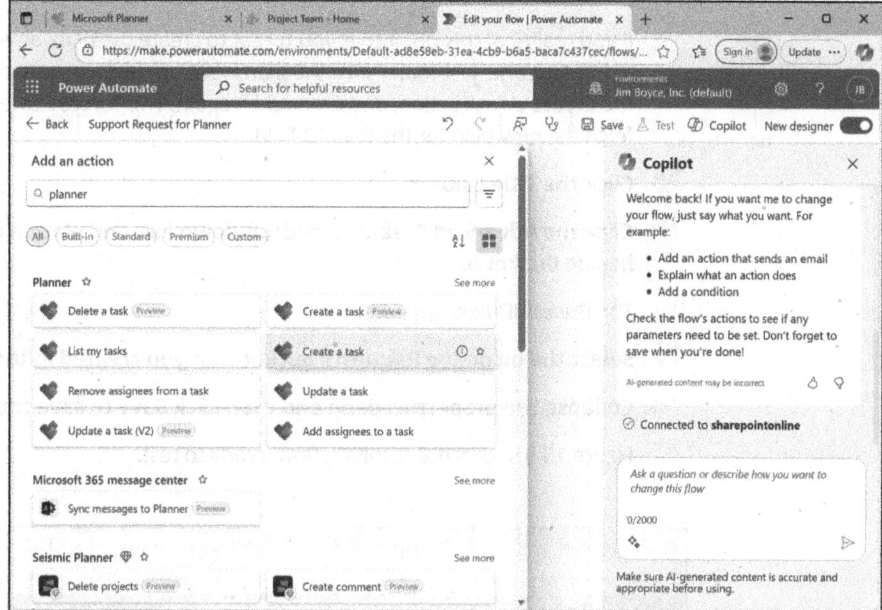

FIGURE 13-7: Choose the Create a Task action.

14. In the Title field, type / and choose Insert Dynamic Content.

A right-side pane appears that contains the columns from the SharePoint list (see Figure 13-8).

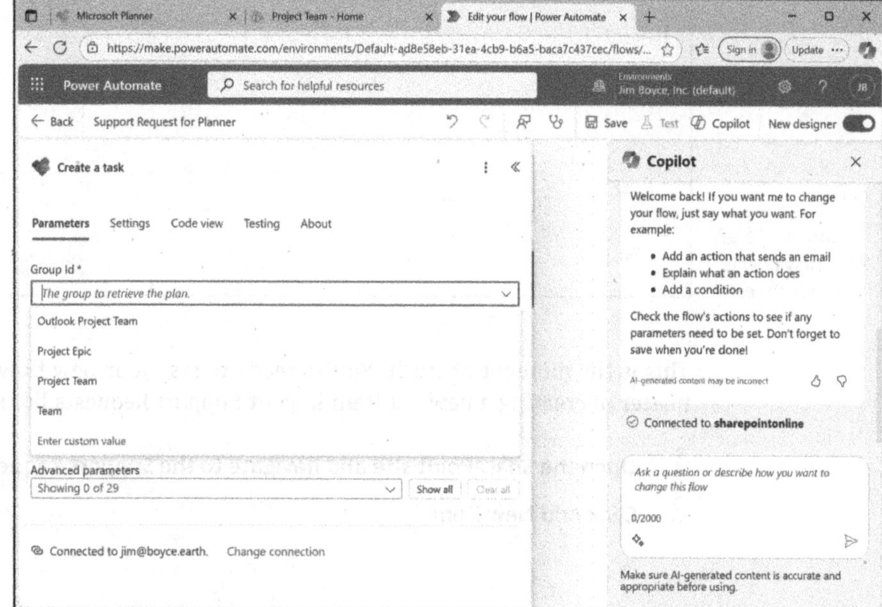

FIGURE 13-8: Enter information for the Microsoft 365 group, the plan, and the SharePoint field to be used for the task title.

TIP

If you realize at this point that you forgot to create the plan, or you want to use a different one, just make sure the plan is set up the way you want it. Delete the Create a Task action, and then add it again. This forces Power Automate to fetch the new plan for the Plan ID field.

15. **Click the Title column.**

16. **Click the Advanced Parameters drop-down list, and place a check mark beside Bucket ID.**

    The Bucket ID field appears.

17. **Select the Incoming Requests bucket that you created in the plan.**

18. **Collapse the properties pane and then click Save to save the flow.**

    Figure 13-9 shows the resulting flow, ready to test.

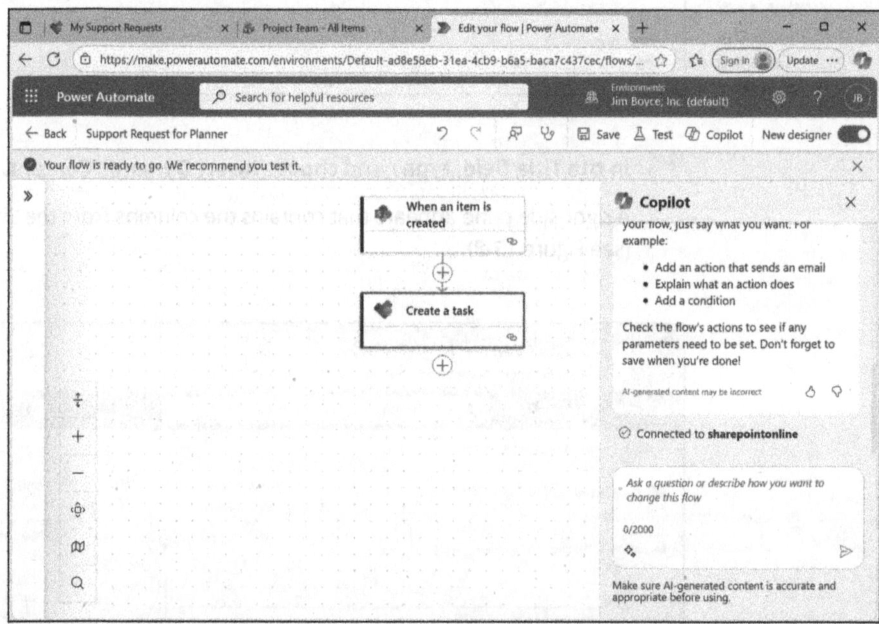

FIGURE 13-9:
Your new flow is ready to test.

This is the moment of truth! You're ready to test your new flow. That's a simple matter of creating a new list item in your Support Requests list in SharePoint:

1. **Open the SharePoint site and navigate to the Support Request list.**

2. **Click Add New Item.**

3. **Enter information for the new item, as shown in Figure 13-10, and click Save.**

The new task should show up in the Incoming Requests bucket in Planner after a short delay. If you don't see the task, refresh the plan window.

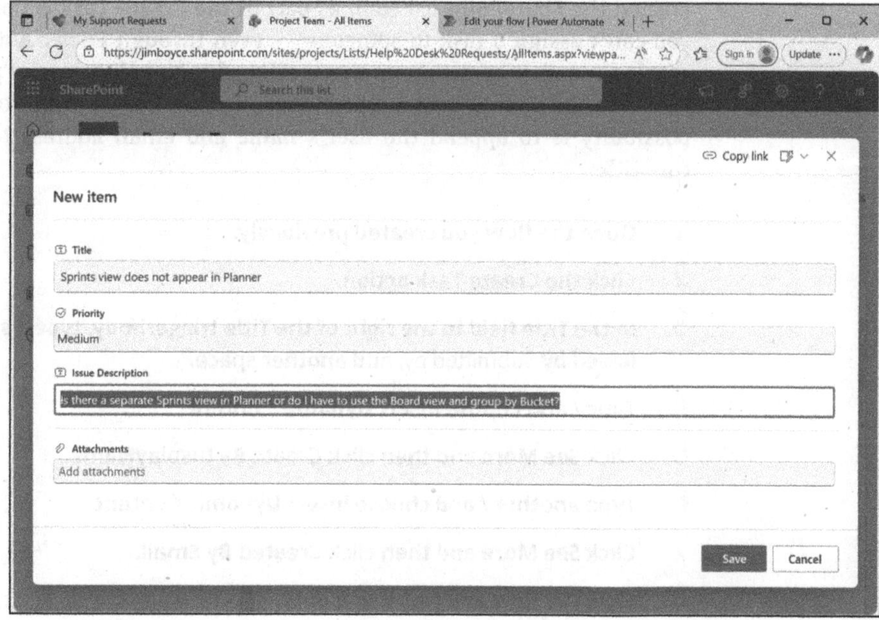

FIGURE 13-10: Create a new support request item in the SharePoint list.

Although this example takes a few minutes to run through, hopefully you've experienced how easy it is to create flows with Power Automate. You could flesh out this flow by adding fields like Technology to the SharePoint list and have that added to the task.

TIP

The default behavior is for SharePoint to add a list to the Quick Launch pane when you create the list. If this didn't happen for some reason, just open properties for the list, click List Name, Description, and Navigation, and choose Yes under Display This List on the Quick Launch.

Recognizing the limitations and tweaks

When you open the task's Details pane after the flow has created the task, you may notice that a name is shown at the bottom of the pane. In the previous example, the pane shows New Task "Sprints View Does Not Appear in Planner" Created. The name shown above this note is the identity that was used to run the flow, not the name of person who created the SharePoint list item. So, how do you capture in the task the name of the person who created the SharePoint list item and needs tech support?

You can't create custom fields in Planner, so you're constrained to using the existing available fields. In this example, you could assign the task to the user who created the SharePoint list item as part of the Create Task action. Unfortunately, you can't assign a task to someone — even through Power Automate — if they aren't a member of the group that's aligned to the plan. You're probably limiting access to the plan to your support team, so that isn't a great solution. One more possibility is to append the user's name and email address to the end of the case title:

1. **Open the flow you created previously.**

2. **Click the Create Task action.**

3. **In the Title field to the right of the Title triggerbody, type a space followed by** Submitted by, **and another space.**

4. **Type / and choose Insert Dynamic Content.**

5. **Click See More and then click Create By DisplayName.**

6. **Type another / and choose Insert Dynamic Content.**

7. **Click See More and then click Created By Email.**

8. **Save the flow and test it again.**

In this example, you should end up with a task that concatenates the title, the user's display name, and their email address, as shown in Figure 13-11.

Also be aware that there are limitations with Planner Basic and Planner Premium plans and their interactions with Power Automate. Planner Basic is fully supported by the Planner connector in Power Automate, which means you have a full range of actions you can perform with Basic plans. These include creating and updating tasks, assigning tasks, adding checklist items, moving tasks to buckets, marking tasks as complete, and triggering flows when tasks are created or completed.

You have some capabilities with Premium plans and Power Automate, limited primarily to reading tasks and triggering flows based on task changes. You can interact with Premium plans using the Microsoft Graph application programming interface (API), but doing so requires custom Hypertext Transfer Protocol (HTTP) calls, app registration in Microsoft Entra ID, and appropriate permissions, all of which are beyond the scope of *Microsoft Planner For Dummies*.

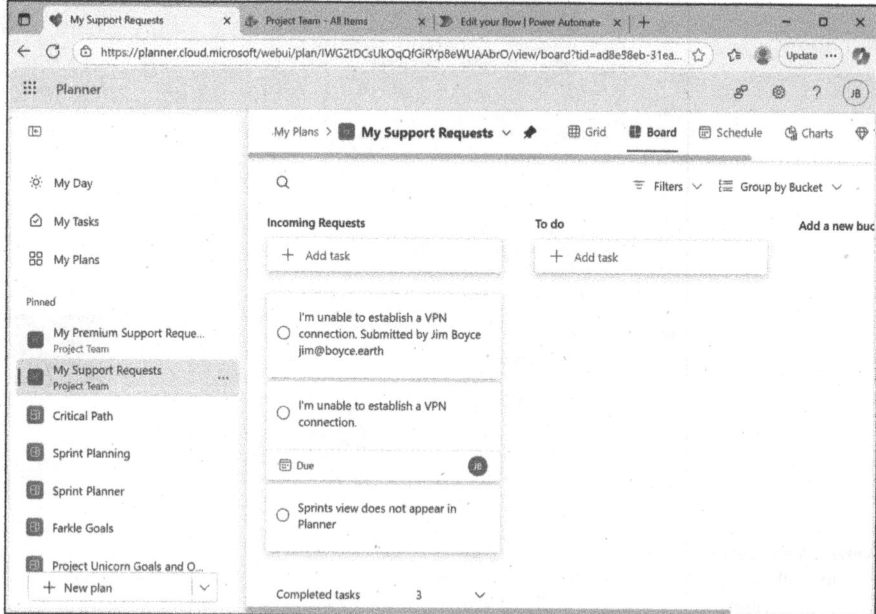

FIGURE 13-11:
The updated flow uses multiple fields from the SharePoint list for the task title.

Connecting Planner with Outlook, To Do, and SharePoint

Planner integrates with several of the Microsoft 365 applications and services, adding collaboration features and enabling you to view and work with Planner tasks and views outside of the Planner web interface or Teams. This chapter explores three: Outlook, To Do, and SharePoint.

Visualizing Planner in Outlook

If you're like most people who use Microsoft 365, you spend a lot of time each day in Outlook. You can integrate Planner with Outlook in a handful of ways, depending on whether you're working with a Basic plan or a Premium plan. I'll start with Basic plans, which offer the most integration.

The first step you may want to take is to publish a Basic plan to an Outlook Calendar. Doing so makes the calendar available as an iCalendar feed so that you and others can subscribe to the feed. You can then view tasks from the project plan in Outlook's Calendar view. The tasks show up as all-day events on the task's due date. You can view the task title, due date, plan name, and a link to the task in Planner. Figure 13-12 shows an example.

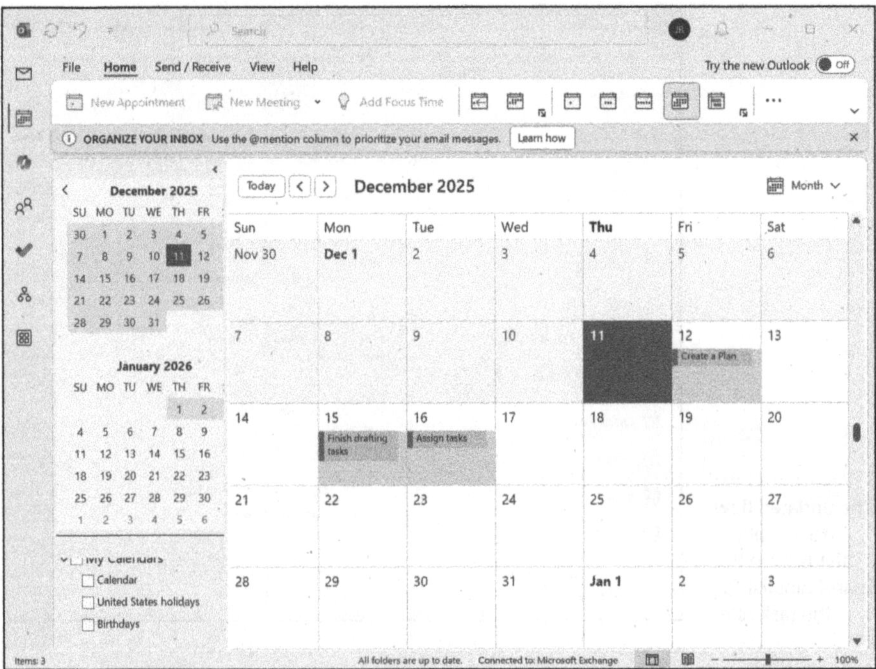

FIGURE 13-12:
View a plan's task in Outlook by subscribing to the plan's iCalendar feed.

WARNING

There is one potentially very big problem with publishing a plan as an iCalendar. Anyone with internet access can subscribe to the feed if they know the URL. iCalendar feeds are public and anonymous, so there is no way to restrict access to only your organization. Consider carefully whether this option is secure enough for your situation.

If publishing a plan as an iCalendar doesn't pose security risks for your organization, you can follow these steps to publish a plan:

1. **Open the Basic plan you want to publish.**

2. **Click the drop-down beside the plan name and choose Add Plan to Outlook Calendar.**

3. **Choose Publish, Share with Anyone (see Figure 13-13).**

4. **Copy the iCalendar link to your clipboard.**

5. **Click Add to Outlook.**

   Outlook opens automatically and gives you the option to Import the feed (see Figure 13-14).

6. **Choose color or other options and then click Import.**

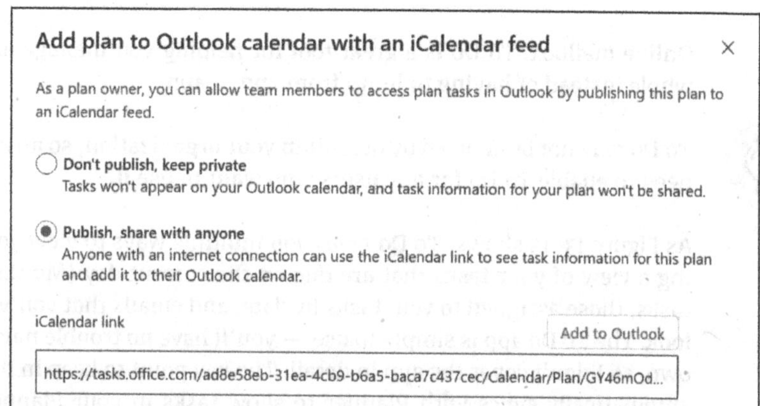

FIGURE 13-13: Choose the option to share the iCalendar feed for others to view.

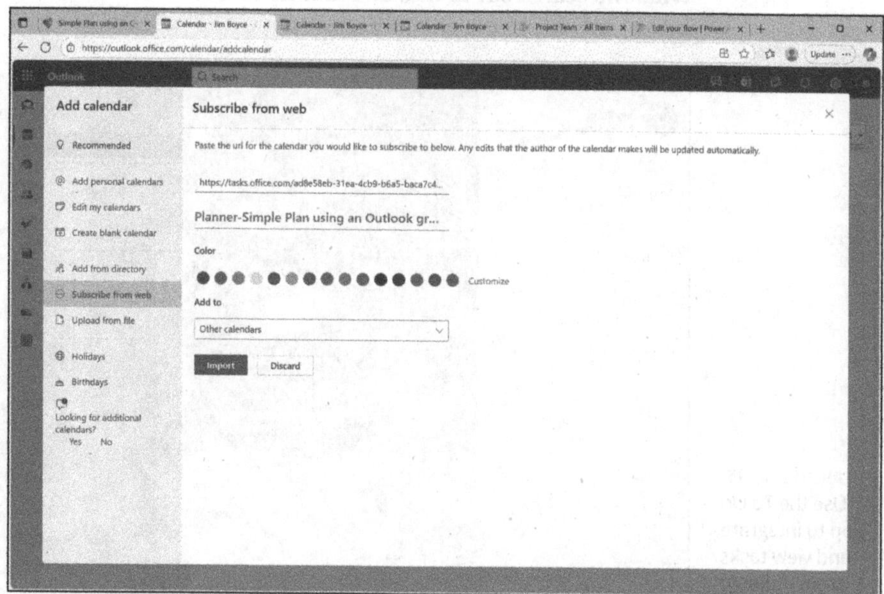

FIGURE 13-14: Choose options for the shared calendar before importing it.

Planner also integrates with Outlook through notifications. If you create an Outlook-enabled Microsoft 365 group, as explained in Chapter 4, you and others will receive email notifications when they're assigned a task, comments are updated, and group updates occur.

Using Planner with To Do

To Do is a simple application that can bring together tasks from multiple sources including Planner, Outlook flagged emails, Outlook tasks, tasks created in To Do itself, and potentially other apps that write tasks to your Microsoft Exchange

Online mailbox. To Do is a great tool for helping you manage all your tasks as a whole instead of having to jump from app to app.

To Do may not be enabled by default in your organization, so an administrator will need to enable To Do for any users who want to use it.

As Figure 13-15 shows, To Do gives you multiple ways to view your tasks, including a view of your tasks that are due on the current day (My Day), high-priority tasks, those assigned to you, tasks by date, and emails that you've flagged in Outlook. The To Do app is simple to use — you'll have no trouble navigating it on your own, so I don't cover the app in detail. The key point to keep in mind is that To Do automatically syncs with Planner to show tasks in your Planner plans that are assigned to you. To Do works with Basic and Premium plans. Just search in your Windows search bar to find and launch To Do.

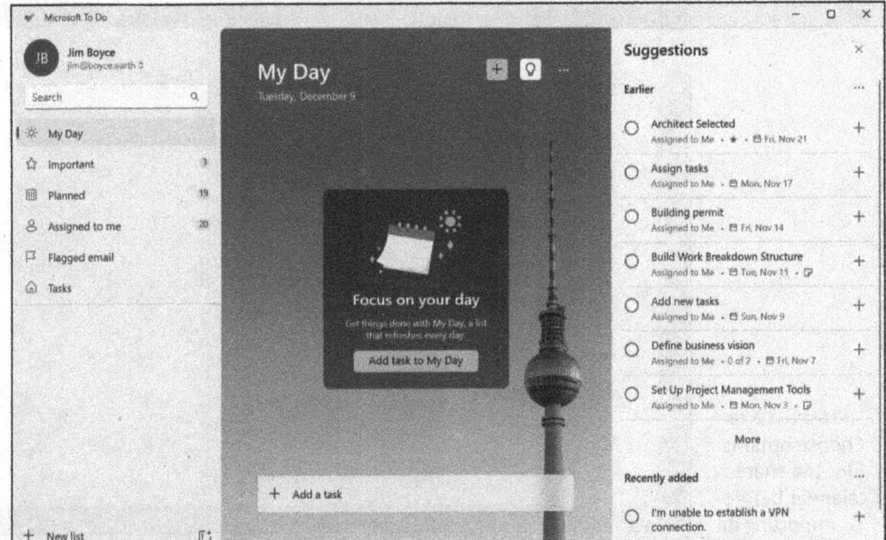

FIGURE 13-15:
Use the To Do app to integrate and view tasks from multiple sources.

Integrating Planner and SharePoint

Each Microsoft 365 Group includes a SharePoint Online site that includes a document library for storing and sharing documents. The site also includes a shared Microsoft OneNote notebook that you can use to share notes, create an informal wiki, and store other information shared by the Group. Because Planner is also a shared component for the Group (assuming you've linked the plan to a Microsoft 365 Group), you can integrate Planner and SharePoint to collaborate with others in the following ways:

- **» Share documents.** You can upload documents to SharePoint document libraries to make them available to others on your project team. You can assign metadata to the documents to identify specific document types, like proposals, bids, quotes, progress reports, and so on. Files that you attach to a task are also stored in SharePoint.

- **» Share a notebook.** The shared OneNote in a SharePoint site is handy for capturing meeting notes, sharing processes, and sharing other types of information related to your project.

- **» Use SharePoint pages to share information via web pages.** Create pages on your SharePoint site to collate information, inform users, host forms, and share information as your team's intranet site.

- **» Use SharePoint lists to capture and share information.** SharePoint lists are a great option for capturing and sharing information in list form. Consider a SharePoint list as an alternative to using a Microsoft Excel file to share data.

- **» Use Power Automate to move information bidirectionally between Planner and SharePoint.** The "Automating Workflows with Power Automate" section, earlier in this chapter, explores ways to capture information in SharePoint lists and create tasks based on those list items. Data can flow from Planner to SharePoint as well, for example updating a SharePoint list item when a task is marked as completed.

- **» Embed Planner in SharePoint pages.** Use the Planner web part to embed a plan in a SharePoint page. Team members can work with tasks in the web part just as they can in the Planner web app.

TECHNICAL STUFF

SharePoint web parts are modular components that you can add to SharePoint pages. Web parts have specific functions like embedding an image, creating scrolling news readers, and much more. The Planner web part enables you to integrate Planner plans directly into a SharePoint page.

Using Planner and Loop

Loop is a cloud-based productivity app that enables users to collaborate in shared spaces for document sharing, taking notes, and simple project planning and task management. Loop is integrated with Microsoft 365, which means it integrates with applications like Excel, Microsoft Word, Outlook, OneNote, Teams, SharePoint, Copilot, Microsoft Whiteboard, and of course, Planner. The following sections explain how Loop can benefit your team and how to integrate Planner and Loop.

Understanding Loop

Loop provides a workspace for teams to collaborate on projects and initiatives. At Microsoft, I worked with people on my team and across other teams and roles on various projects and initiatives. Loop was an important element that enabled everyone to share information, plan activities, take feedback, coedit documents, capture meeting notes, and more.

For example, if I needed others on the team to review or add content, I just added the Loop component to a Teams chat, and everyone could click through to work on that content. If my manager needed the team to capture key metrics, recurring themes for support issues, or other data, she could create a Loop component and share it with the team through a Teams chat and a follow-up email; then the team could just click through to add their own information.

To truly take advantage of Loop, you need to understand its parts. Loop is composed of three main elements:

- **Loop workspaces:** Workspaces are high-level containers in Loop that bring together Loop components, pages, files, tasks, and project resources. You can think of a Loop workspace as a project hub for your team.

- **Loop components:** Components are live-content elements that you can embed and use in various ways, such as in emails, Teams chats, Loop pages, and other places. Loop components are live elements, so they stay synchronized wherever you find them. A Loop component in a Teams chat, for example, will show the same content in an email. Change the content, and the change synchronizes everywhere it appears.

- **Loop pages:** Think of a Loop page as a canvas where you can add text, images, and links; insert Loop components; and embed Planner task lists. You can also use Copilot to summarize your data and generate content. Your Loop project hub may have a single page, or it may have multiple pages for different aspects of the project like planning, meetings, reference resources, and more.

Creating a Loop workspace

The first step to take in using Loop with Planner (or by itself) is to create the Loop workspace. Although you can use Loop components outside of a workspace, creating the workspace and adding components to it will help you experience and understand Loop.

Follow these steps to create a Loop workspace:

1. **Open a browser and navigate to** `https://loop.cloud.microsoft`.

2. **Click the plus (+) button at the top of the left navigation pane and choose New Workspace.**

   The Create a New Workspace dialog box appears.

3. **Type a name for the workspace in the Type a Name field.**

4. **Choose a sensitivity label, if needed.**

5. **Click the Invite Members with Name or Email field and add users as needed to the workspace.**

6. **Click Create.**

   Figure 13-16 shows the resulting workspace.

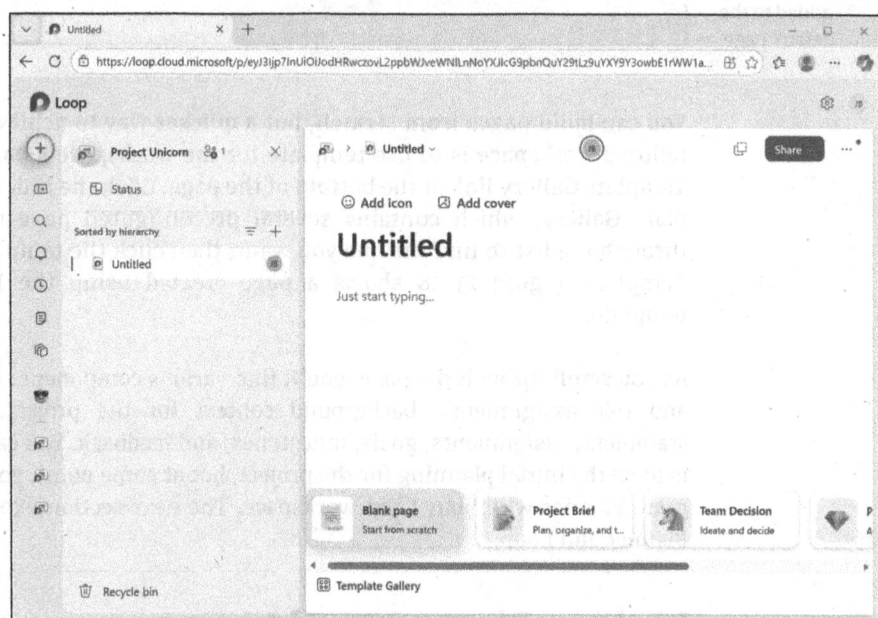

FIGURE 13-16: A Loop workspace named Project Unicorn.

The workspace starts with a blank Loop page, and it's easy to modify it. For example, just double-click the title (which starts out as Untitled) and type the title. Click Add Cover to choose a graphic for the top of the page. Click Add Icon if you want to add an icon at the bottom-left corner of the cover image. You can also just start typing to add text to the page. Figure 13-17 shows an example of my Project Kickoff page with a cover added.

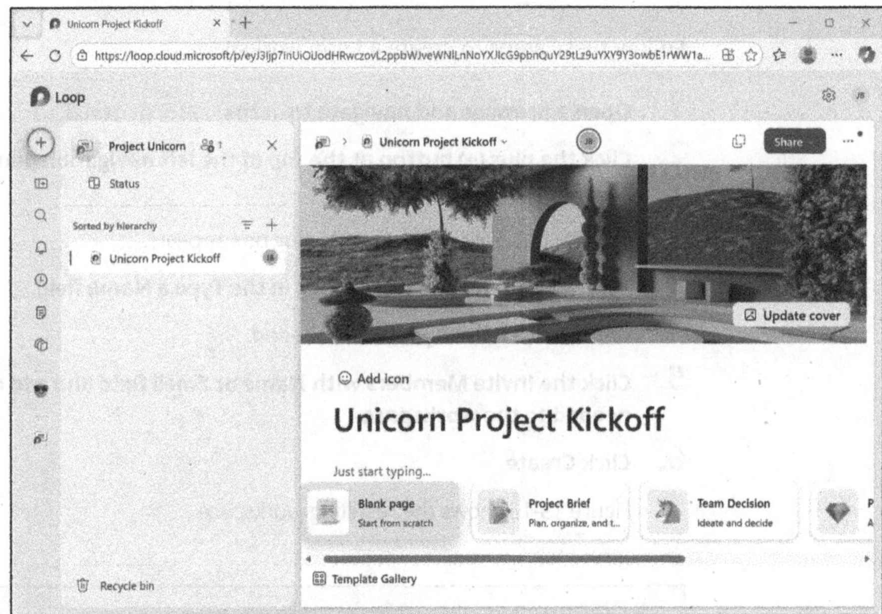

FIGURE 13-17: A title and cover added to the default page.

You can build pages from scratch, but a quicker way to achieve a professional, tailored workspace is to use template for the workspace's pages. You'll find a Template Gallery link at the bottom of the page. Click the link to open the Template Gallery, which contains several preconfigured page templates. Scroll through the list to find the one you want; then click the template and click Use Template. Figure 13-18 shows a page created using the Project Planning template.

As you scroll through the page, you'll find various components that capture roles and role assignments, background context for the project, an opportunity statement, assignments, goals, milestones, and feedback. You can use the page as is to do the initial planning for the project, but at some point, you'll want to integrate Planner with your Loop workspace. The next section explores how to link Planner and Loop.

Linking Planner and Loop

Planner Basic and Planner Premium are on different development paths. As such, some features are available (as I write this) for Basic plans but not for Premium plans because those capabilities are still in development for Planner Premium versions. Loop integration is one capability that works for Basic plans and not for Premium plans; this will likely be added through future development, but it isn't available now.

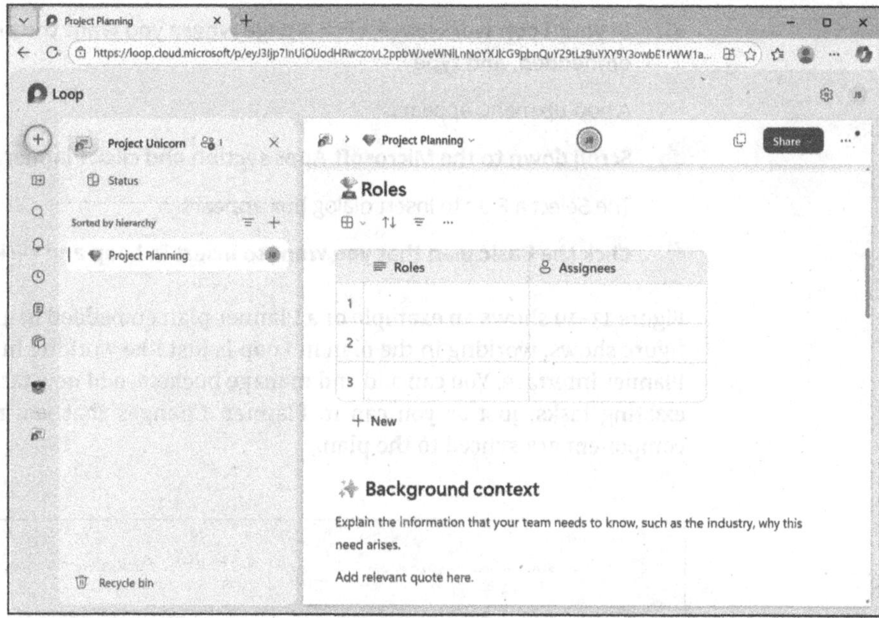

FIGURE 13-18: This page uses the Project Planning template.

Embedding a Planner plan in Loop is easy, but before you start the process, make sure you've satisfied these requirements:

>> You have a Microsoft 365 Group, and it has a SharePoint site.

>> The group is not hidden.

>> Your admin has enabled Loop and SharePoint storage for loop.

>> You're embedding a Basic plan.

The easiest way to ensure that you have a Group set up that will work with Loop is to create the Group from Teams. To do that, open Chat in Teams, click the **ellipsis** at the top of the Chat pane, and choose Your Teams and Channels. Click Create Team and work through the Create a Team page to create the team.

Follow these steps to link a Planner Basic plan to the workspace:

1. **Create a Basic plan in Planner and add some buckets and tasks.**

    You can add more items to the plan later, if needed.

2. **In your Loop workspace, click a page where you want the plan to be embedded, and type /.**

    A pop-up menu appears.

3. **Scroll down to the Microsoft Apps section and click Planner.**

    The Select a Plan to Insert dialog box appears.

4. **Click the Basic plan that you want to insert in Loop and click Insert.**

Figure 13-19 shows an example of a Planner plan embedded in a Loop page. As the figure shows, working in the plan in Loop is just like working in Board view in the Planner interface. You can add and manage buckets, add new tasks, and work with existing tasks, just as you can in Planner. Changes that you make in the Loop component are synced to the plan.

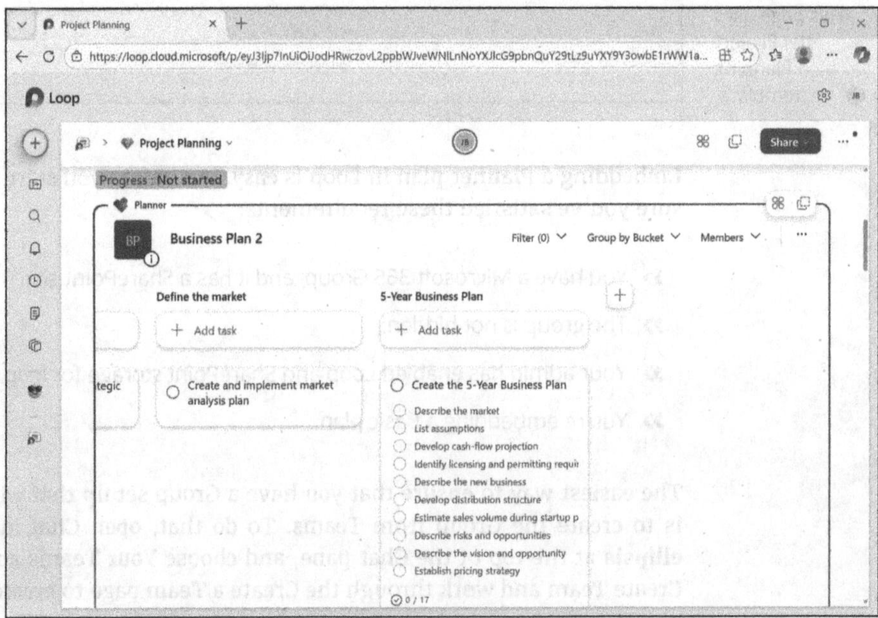

FIGURE 13-19: A Basic plan added to a Loop page.

You can insert a link to a Premium plan in a Loop page, which enables you to open the plan in the Planner web app. However, the plan will not open inside Loop.

Bringing it all together

You could accomplish results similar to Loop by creating a SharePoint site, customizing some pages, and embedding a plan on a page with the Planner web part. But that's a much more complex task, particularly if you're not very familiar with creating and managing SharePoint sites and pages.

Where Loop shines is in its simplicity and ease of use. It's easy to build a project hub using a Loop workspace and integrate documentation, forms, notes, feedback, and lots more, including Planner plans. Best of all, wherever you insert Loop components — whether Teams, Outlook, or other locations — those components will always be up to date. All the features come together to give you a pervasive way to collaborate with others across a broad range of apps and work styles.

Chapter **14**

Integrating Copilot with Microsoft Planner

Microsoft Copilot is a productivity assistant powered by artificial intelligence (AI), built into Microsoft 365. Copilot is integrated into several Microsoft 365 applications like Microsoft Outlook, Microsoft Word, Microsoft Excel, Microsoft PowerPoint, Microsoft Teams, Microsoft Planner, and more. Copilot can boost productivity by completing tasks for you, automate repetitive tasks, provide contextual intelligence by analyzing your content to provide insights and recommendations, and generate content like emails and presentations.

This chapter explores Copilot to help you understand how to license it and put it to work across your application stack, including in Planner.

Getting Ready for Copilot

AI isn't a new concept, but it *is* experiencing an unprecedented surge in development, implementation, and adoption across many aspects of our daily lives. AI is transforming how we search for and analyze information, complete work tasks, collaborate with others, and interact with systems and services.

Although AI is certainly transforming our personal and work lives, it isn't all unicorns and rainbows. For example, AI requires massive computing power, which means more data centers and more power and other utilities to drive those data centers. These requirements are putting a strain on power grids and water supplies and even causing municipalities to take a "not in my backyard" position on locating data centers in their areas. AI is also putting pressure on many technology companies to try to keep up with the competition, which often means refocusing their businesses, cutting staff to free up capital for investment, and making other changes to try to adapt. So, although tools like Copilot can transform the way we work, they'll also transform our society in other ways — and not always for the good.

Many AI offerings are available today. Because of Copilot's integration with Planner, Copilot is the AI tool this book focuses on. The following section explores what Copilot is and how it works.

Introducing Copilot and what it can do for you

Copilot is a family of AI-powered assistants that are integrated into the Microsoft 365 ecosystem. These assistants enable you to use natural-language queries to perform a wide range of tasks within your Microsoft 365 applications. Copilot assistants combine the following:

>> **Large language models (LLMs):** LLMs can interpret natural-language queries and generate humanlike responses to summarize data and present content, provide recommendations, draft content, and more. LLMs are trained by parsing vast amounts of data. They can perform seemingly humanlike tasks, like predicting the next word in a sentence, understanding context, generating humanlike content, and answering questions based on the data in the model.

>> **Microsoft Graph:** Copilot can incorporate your organization's data, including emails, documents, meeting chats and transcripts, files, and lots more. Because it has access to not only your organization's data but also your own data, Copilot can analyze and answer queries about your email, your

"

meetings, company processes, and much more, integrating company data
with your own.

>> **Microsoft 365 applications:** Copilot is embedded in and works directly with
many Microsoft 365 applications. These include, but are not limited to,
Outlook, Word, Excel, PowerPoint, Teams, and Planner. This enables Copilot to
analyze that data, combine it with other information stored in your organizational data layer when needed, and create content for you. It can also provide
summaries and answer questions about that content by providing it with
natural-language queries like, "Summarize all the emails in my Customer
folder from the past three months, organized by customer, technology, and
customer sentiment."

Just because Copilot or other AI tools can answer your questions doesn't mean the
answer is right. Many times, I've asked Copilot to research a topic and provide a
summary, and at least some of the resulting information is simply wrong. This
situation can be caused by incomplete or inaccurate data in the model, biased
data, ambiguous or unclear input (AI misunderstood your question), or what are
called *hallucinations* (in which AI just makes things up). Keep in mind that, despite
its name, AI is not *actually* intelligent. It's generally very good at what it does, but
it's not infallible.

Following are some suggested uses for Copilot across some of the most frequently
used Microsoft 365 applications. Although these aren't directly relevant to Planner, they'll help you understand Copilot's capabilities more deeply:

>> **Outlook:** Summarize your emails to find topics associated with specific
senders or companies. Search for emails relating to specific technologies,
products, topics, or projects. Interpret a sender's sentiment in an email
thread. Draft an email for a topic based on previous emails, new information
in your organization's data layer, external resources, and more.

>> **Excel:** Summarize the data in a table and identify trends in the data.
Determine top products by quantities or revenue. Suggest why values are
increasing or decreasing for sales, revenue, or other factors. Remove dupli-
cates in your data, split a full-name column into first and last names. Create a
chart summarizing the data in a table by simply telling Copilot what type of
chart you want.

>> **Word:** Copilot can do a lot more in Word than just guess the next word or
phrase in a sentence that you're typing. It can use various sources and
references to draft documents and reports, create an article, clean up your
grammar or prose, or simplify complex content. Bear in mind that the results
will still require editing for accuracy and content.

>> **PowerPoint:** Insert charts or graphs based on external data, recommend and implement changes to improve a presentation, generate custom slide backgrounds, create images, or even create an entire presentation from a single natural-language prompt. For example, you can create an interesting presentation with the following prompt: "Create an eight-slide presentation on the biology, habitat, and lifespan of the platypus, and include images."

>> **Teams:** Summarize a meeting with participants, key points, and action items. Automatically extract and assign tasks from meetings. Summarize long chat threads. Summarize conversations and convert them to documents. Catch up on a part of the meeting that you missed.

>> **Planner:** Use natural-language prompts to create plans and tasks. Extract task lists from a variety of sources and convert them to tasks in Planner. Summarize a plan. Provide status updates. Identify dependencies, risks, overdue items, and resource conflicts. Break major tasks into subtasks. Identify critical-path tasks. Generate executive summaries, weekly status reports, and other reports.

Identifying the requirements and licensing for Copilot

The requirements necessary to use Copilot are tied mostly to how you license it. Until August 2025, Planner Premium users could leverage Copilot with no additional cost. Microsoft changed licensing requirements, and Copilot is no longer available with Planner licenses. To use Copilot with Planner, you must have one of the following Copilot licenses:

>> **Microsoft 365 Copilot Business:** This license adds Copilot to Word, Excel, PowerPoint, Outlook, Teams, Planner, Microsoft Loop, Microsoft SharePoint, and Microsoft OneDrive. It can be used with Microsoft 365 Business Basic, Business Standard, and Premium.

>> **Microsoft 365 Copilot Enterprise:** This license adds Copilot to the same set of applications and is for use with Microsoft 365 E3, E5, Business Standard (No Teams) + Copilot, and Business Premium.

In addition to these licensing requirements, you must have a Planner Premium subscription and work through the Planner web app or in Teams.

Microsoft rolls out features in a staggered fashion across *tenants* (separate and secure instances of Microsoft's cloud offering). Because these rollouts don't happen across all tenants at the same time, not everyone gets new or updated features at the same time. For example, as I write this, I have Copilot capabilities in the

Planner web app, but Copilot was conspicuously missing from Teams for me — but it showed up a day later. If you're sure your licensing is correct but you're not seeing a feature that you expect to see, it could be that the feature hasn't rolled out to your tenant yet. You can open a support case with Microsoft to check.

Other Copilot license options are available, but the requirements described here are the ones that enable Copilot in Planner Premium subscriptions and plans. Note that licensing changed shortly before I wrote this and will likely change again in the future. Check the latest Microsoft online licensing sources to confirm licensing requirements if you need to add Copilot capabilities to Planner. The most direct source is `www.microsoft.com/en-us/microsoft-365/planner/microsoft-planner-plans-and-pricing`.

Simplifying Tasks with Project Manager Agent

Copilot can simplify and automate several actions in Planner, ranging from something as simple as creating a new plan to much more complex activities like analyzing a plan and providing a list of critical-path items. The following sections explain how to use Copilot with Planner Premium plans to perform common plan activities.

Finding Copilot in Planner

I confess to considerable confusion when I first started looking for Copilot in Planner in preparation for writing this chapter. Copilot was nowhere to be found. The first problem was that the licensing had changed shortly before, and the Copilot features that were there before were gone. I fixed the licensing issue, and there was still no Copilot icon! But, lo and behold, a new thing popped up called Project Manager agent. Figure 14-1 shows its icon in Teams (the same icon also shows up in the Planner web app when you open a Premium plan). So, Project Manager agent is the new Copilot experience for Planner, replacing the previous Copilot in Planner experience.

You'll also see a Copilot icon in the Planner web app, but this is Microsoft Edge Copilot. Although you can perform some Planner-related activities with it, this chapter focuses on Project Manager agent.

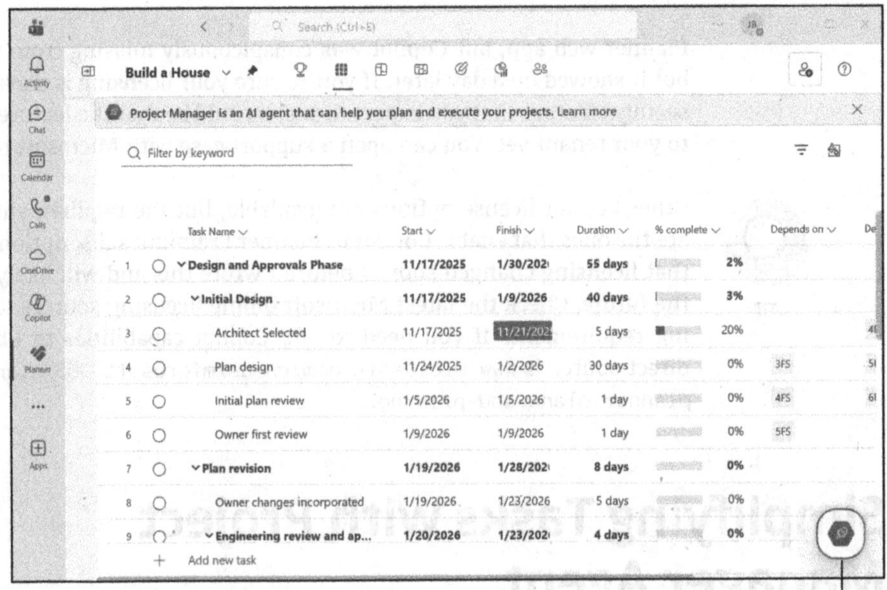

FIGURE 14-1:
Project Manager agent is your ticket to Copilot in Planner.

Project Manager agent icon

Understanding Project Manager agent's capabilities

Project Manager agent offers an impressive range of capabilities that are not limited to just offering advice or creating tasks. Its capabilities go much deeper due to the combined integration with Planner, your organizational data, and external web resources. Following is a summary of its key capabilities:

>> **Plan creation:** The agent is exposed within your Premium plans, so you don't use the agent to create plans per se, but rather to flesh them out after you create them. You can create a blank Project Manager agent plan or use a template. Then you can start using the agent to flesh out the plan.

>> **Task creation:** You can use the agent to create tasks in the plan by giving it a list of tasks, but the agent can operate at a much higher level. For example, you can create a goal in the plan, add some supporting documents for the goal, and then ask the agent to generate tasks based on the goal. The agent creates tasks based on that data, and you can add, modify, or delete tasks as needed afterward.

>> **Task execution:** In addition to creating tasks with Project Manager agent, you can also have it *execute* tasks. The process is logical: You assign a task to the agent, and it begins to work on the task. You can assign multiple tasks to the agent, which it queues and then works on. The agent uses four states to indicate progress: Queued, In Progress, Needs Input, or Done. If the agent

needs more input on a task, you provide the input and then regenerate the task to cause the agent to continue working on it.

>> **Integrate web resources:** Project Manager agent can use external web resources, which enables it to provide recommendations, offer guidance, or create tasks using a very broad base of knowledge. It can potentially take into account best practices, industry standards, and other criteria to supplement the data that you provide directly through prompts or supporting documentation.

>> **Reporting:** Project Manager agent can generate reports for you automatically, and you can use natural language to specify the information you'd like to see in the report. The agent then analyzes the plan and generates the requested report.

>> **SharePoint and Loop:** The agent automatically creates a Loop page in SharePoint for every task that it executes. It uses SharePoint to store resource files, giving you another mechanism for interacting with the agent and the data that it creates.

TIP

Covering all the things you can do with Project Manager agent across a broad range of project types would fill a book all on its own. This chapter takes a focused approach with one scenario to help you become familiar with how to work with the agent in a foundational way. From that base, you can begin to leverage Project Manager agent in other scenarios.

Using Project Manager agent templates

Enabling Project Manager agent in Planner adds additional Planner templates. In addition to the Basic and Premium templates, Planner offers a set of Plan with Project Planner templates. Currently, these include the following:

>> **Research Report:** Use this template to gather and analyze data, summarize key findings, generate a cover page, and finalize the report.

>> **Competitive Analysis:** This template helps you identify key competitors and gather information on their offerings, analyze market position and customer base, assess strengths and weaknesses, prepare a report of findings, and present the finalized report.

>> **Learning Path:** Define a topic and identify key concepts and skills, summarize available training resources, break up concepts into consumable chunks, engage in practice exercises, monitor progress and adjust the plan as needed, and summarize learned concepts.

>> **SWOT Analysis:** Perform a strengths, weaknesses, opportunities, and threats (SWOT) analysis. Research strengths and weaknesses to identify opportunities and threats; then summarize and present a report of the findings.

>> **Market Study:** Define scope and objectives, evaluate competitors, evaluate market and demand, conduct a competitive analysis, and compile the findings into a report.

>> **Campaign Launch:** Build and launch a campaign by researching the target audience, analyzing trends, developing a strategy, and then planning and executing the launch.

REMEMBER

The Project Manager agent templates are different from the Basic and Premium templates offered in Planner in that they integrate the agent into the plan and enable the agent to automate much of the work involved in building the plan and executing tasks.

Creating a plan using these templates is essentially the same as starting plans from the other templates:

1. **In Planner, click New Plan.**

   The Create New dialog box appears.

2. **Scroll down and click See All Templates.**

3. **In the Project Manager group of templates (see Figure 14-2), click the one you want to use and click Use Template.**

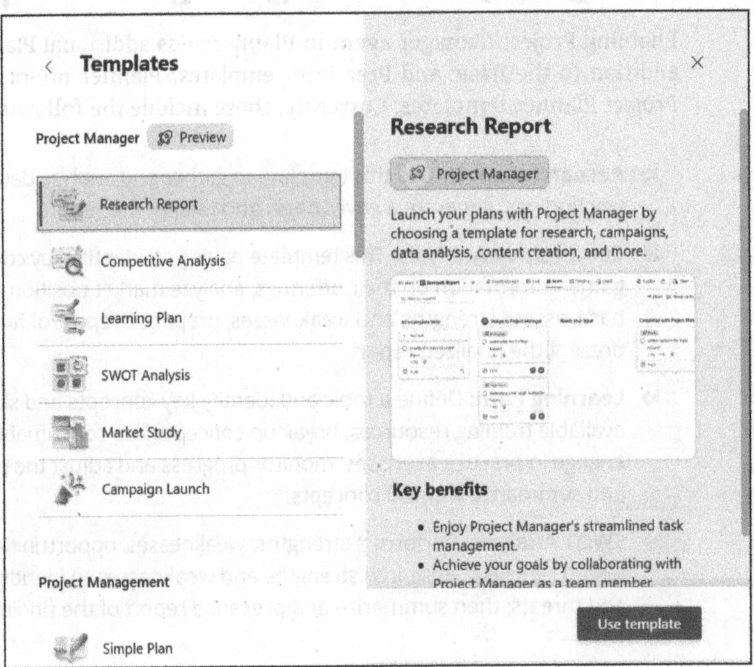

FIGURE 14-2:
Planner offers several Project Manager agent plans.

4. **Type a name for the plan, add it to your pinned plans if you want, and select a group for the plan.**

5. **Click Create.**

One key difference when creating a template for a Project Manager agent template plan is that you must assign a group in order to create the plan.

Creating a plan

The first step in leveraging the power of Project Manager agent is to create a plan. You could use one of the agent templates to create the plan, but let's start a new one from scratch:

1. **Open Planner and click New Plan.**

   The Create New dialog box appears (see Figure 14-3).

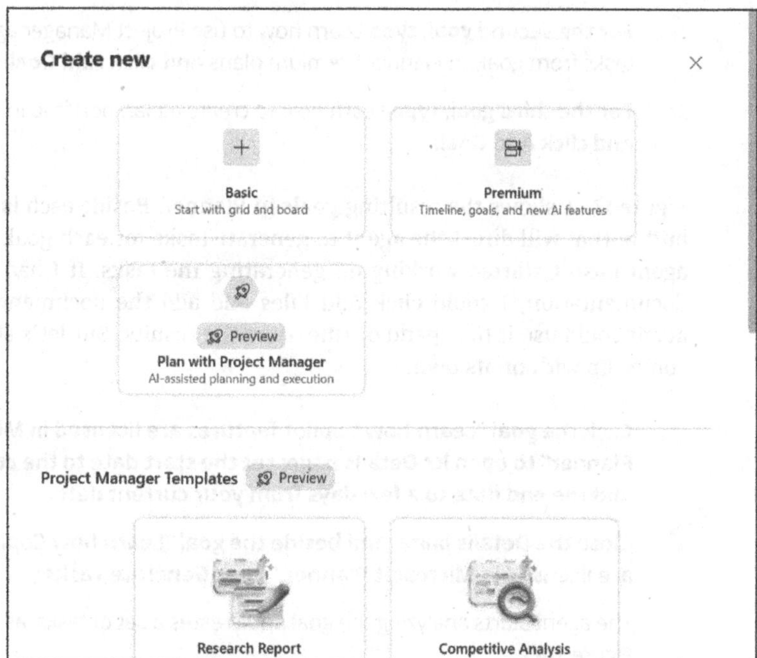

FIGURE 14-3: Start a Project Manager plan from scratch.

2. **Click Plan with Project Manager.**

   The Plan with Project Manager from Scratch dialog box appears.

3. **Type a name for the plan, add it to your pinned plans if you want, and select the group for the plan.**

4. **Click Create.**

Adding goals and supporting resources

To illustrate the analytical capabilities of Project Manager agent, the approach I take in this section is to add some goals and supporting resources, and then direct the agent to create the tasks for the plan. The scenario is simple: I want a plan to learn more about Planner's capabilities!

1. **Use the Project Manager scratch template to create a plan named "Get started with Project Planner" (see the preceding section).**

2. **Open Goals view and click Add Goal.**

3. **For the first goal, type** Learn how Copilot features are licensed in Microsoft Planner **and click Add Goal.**

4. **For the second goal, type** Learn how to use Project Manager agent to create tasks from goals in Planner Premium plans **and click Add Goal.**

5. **For the third goal, type** Learn how to create a plan portfolio in Planner Plan 1 **and click Add Goal.**

Figure 14-4 shows the resulting goals in Planner. Beside each is a Generate Tasks button that will direct the agent to generate tasks for each goal. At this point, the agent hasn't started working on generating the tasks. If I had some supporting documentation, I could click Add Files and add the documentation so that the agent could use it to expand or fine-tune the results. But let's see what the agent comes up with on its own:

1. **Click the goal "Learn how Copilot features are licensed in Microsoft Planner" to open its Details page; set the start date to the current date and the end date to a few days from your current date.**

2. **Close the Details pane, and beside the goal "Learn how Copilot features are licensed in Microsoft Planner," click Generate Tasks.**

   The agent starts analyzing the goal and creates a set of tasks, as shown in Figure 14-5.

3. **Repeat Steps 1 and 2 to have the agent generate tasks for your other two goals.**

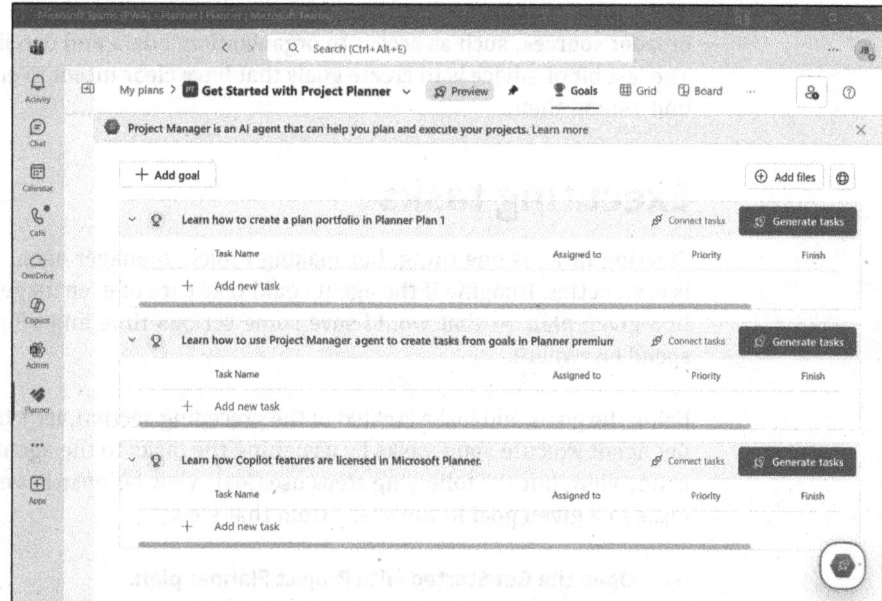

FIGURE 14-4:
Three goals
added to Planner.

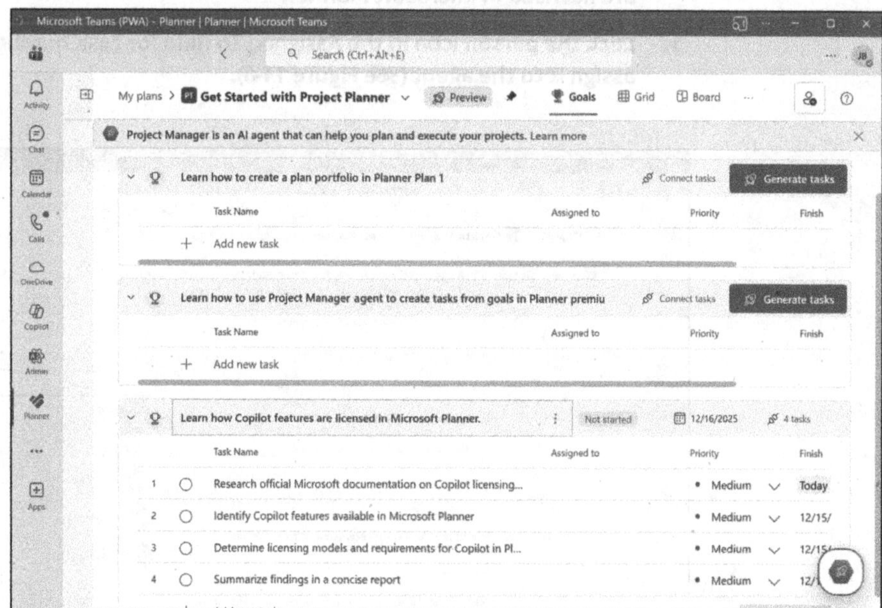

FIGURE 14-5:
Project Manager
agent has created
tasks by analyzing
the goal.

I find the capability that the agent has to create tasks from goals pretty impressive, even when you don't provide a lot of detailed supporting information. The quality and scope of the tasks only improve with additional documentation and broader sources, such as access to organizational data and detailed requirements. The last bit of advice is to create goals that have clear intent to enable the agent to understand them.

Executing tasks

Creating tasks is one thing, but making Project Manager agent *perform* the tasks is even better. Imagine if the agent could take care of even 25 percent of the tasks in a given plan — that would save some serious time and effort that you could spend elsewhere.

Using the goals and tasks created in the preceding section, let's have Project Planner agent execute some tasks by assigning the tasks to the agent. You can do that in any view, but the following steps use Goals view because it's easier to assign all tasks in a given goal to the agent from that view:

1. **Open the Get Started with Project Planner plan.**

2. **Open Goals view and expand the goal named "Learn how Copilot features are licensed in Microsoft Planner."**

3. **Click the person icon in the Assigned To field for task number 1 and assign it to the agent (see Figure 14-6).**

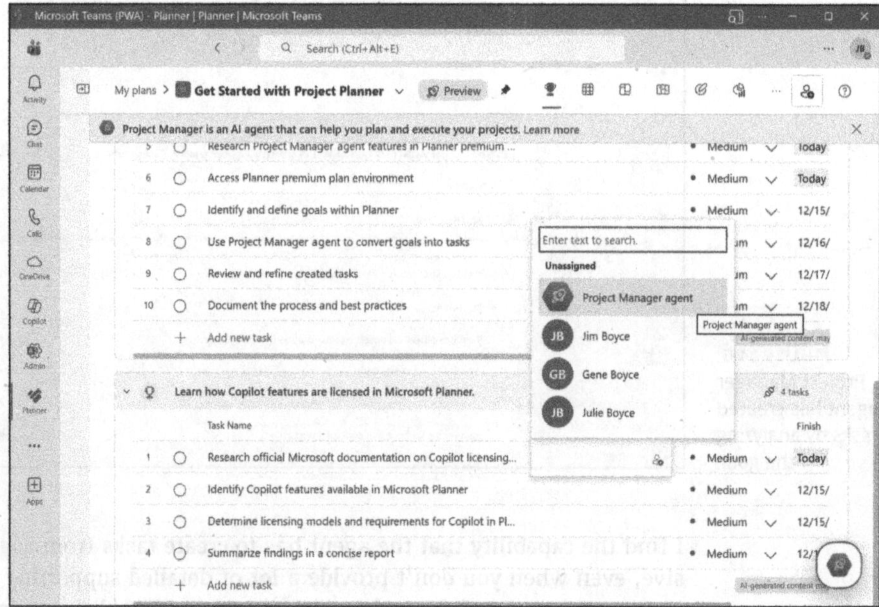

FIGURE 14-6: Assign tasks to Project Manager agent.

4. **Repeat Step 3 to assign all the other tasks in this goal to the agent.**

5. **Switch to Board view, where you can see that the Assign to Project Manager bucket now contains the tasks that you assigned to the agent (see Figure 14-7).**

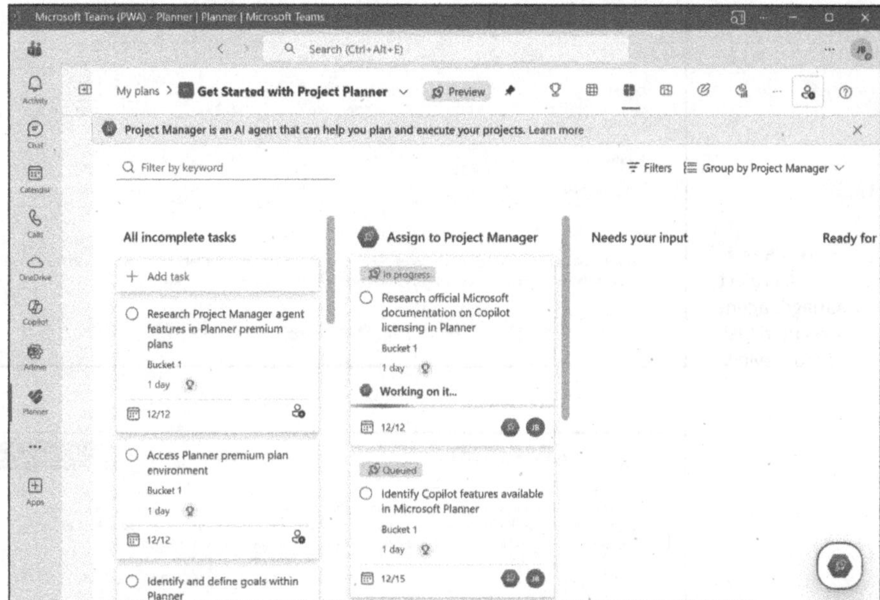

FIGURE 14-7: The tasks have moved to the Assign to Project Manager bucket.

As the agent processes tasks, the tasks will move either to the Needs Your Input bucket or to the Ready for Review bucket. The agent displays a pop-up notification message when a task needs input or review. Or just check the applicable bucket to check progress.

When a task is ready for review, click the task in the Ready for Review bucket. Figure 14-8 shows an example of the task's Details page. Note that the task now includes a green Ready label, indicating that it's ready for review. The agent has also added notes to the task.

Next, scroll down on the task's Details page and note the extensive information that the agent has added to the task (see Figure 14-9). In this example, the agent has researched the licensing requirements and provided a wealth of information about which licenses are required, how to manage and assign the licenses, regional availability of features, a summary of features, costs and add-ons, supporting documentation, future updates, a summary table of requirements, a conclusion statement, and numerous web references with links. Imagine how long it would've taken you to research and develop that same body of information. That's a real time-saver!

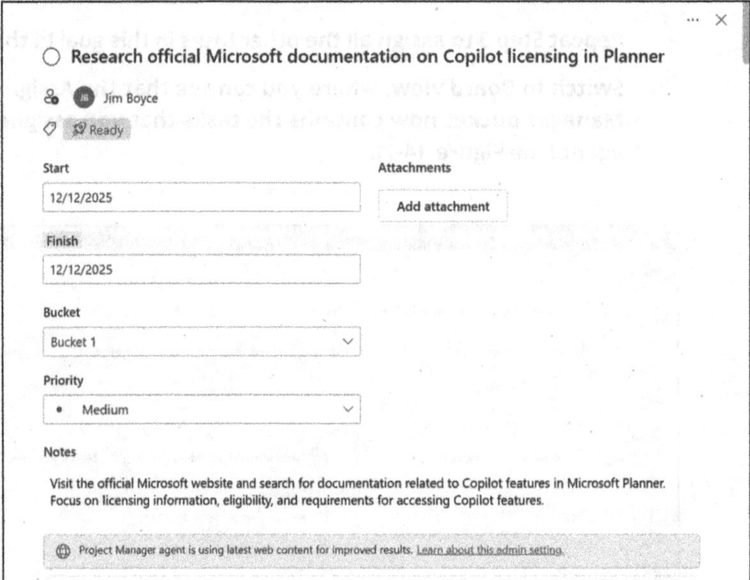

FIGURE 14-8:
A Project Manager agent executed task ready for review.

FIGURE 14-9:
Project Manager agent has compiled a significant amount of supporting information.

Now, what about changes? Let's say you get pulled from this project for a month to work on something critical. Now you've come back to the plan and discovered that you want to add some additional supporting documentation and factor in any changes that may have taken place since you last touched the plan. In that case, add the documentation, refine the goal description, and then click the Regenerate button on the task's Details pane. This moves the task back into the Assign to Project Manager bucket and sets it to either Queued or In Progress. Project Manager agent reprocesses the task and provides you with an update to review.

In some cases, the agent determines that it needs additional input from you to execute the task. When this happens, the agent includes questions directly in the Loop component of the task's Details page. Provide the requested information within the Loop component, and then click Regenerate. This moves the task back to the Assign to Project Manager bucket and queues it for further action from the agent.

As you review an agent-executed task, you'll soon discover that the agent never sets the task to 100 percent completion status. This gives you control over when the tasks are marked as completed so you can control the plan and project flow. So, when you're satisfied with the execution of the task, just mark it completed.

Finding the loop

I mentioned earlier in this chapter that Project Manager agent creates a Loop page for every task that it executes. Although you can work with queries and the other data right in the task's Details page, having it in SharePoint also offers the advantages of organizing the pages in one place and enabling others to view and work with the content without opening the plan or the task.

To find the Loop pages in SharePoint, open the plan and click SharePoint in the plan header. If you don't see SharePoint, click the ellipsis in the head and click SharePoint. A browser opens and shows the Loop pages (see Figure 14-10). Just click a page to open it in Loop. Figure 14-11 shows an example.

Keep in mind that Project Planner agent doesn't create a Loop workspace for a plan. Instead, it creates Loop pages in SharePoint. Although you can view those pages in the Loop interface, they aren't listed in Loop by default because they don't exist in a workspace. You can find them in Loop by searching for them. Just click Search in the navigation pane and type the goal name to find them.

Tracking progress

Keeping track of what's going on in a project is one of the most important activities that drives project success. Planner includes several capabilities for reporting on plan progress, and Chapter 15 explores the Charts view and other methods that you can use to visualize progress and status. Because this chapter focuses on Copilot and Project Manager agent, I cover here the capabilities that Project Manager agent provides for generating reports.

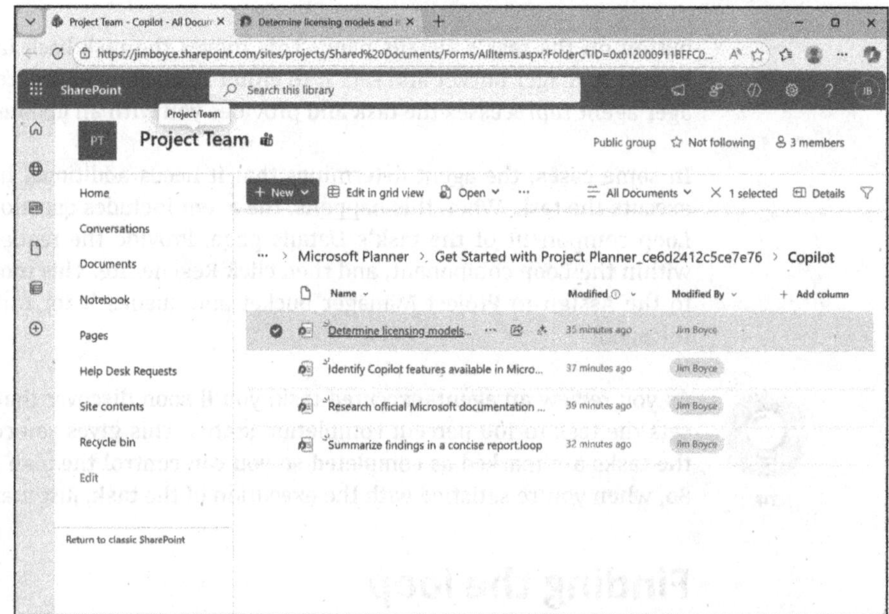

FIGURE 14-10: These Loop pages were created by Project Planner agent in SharePoint.

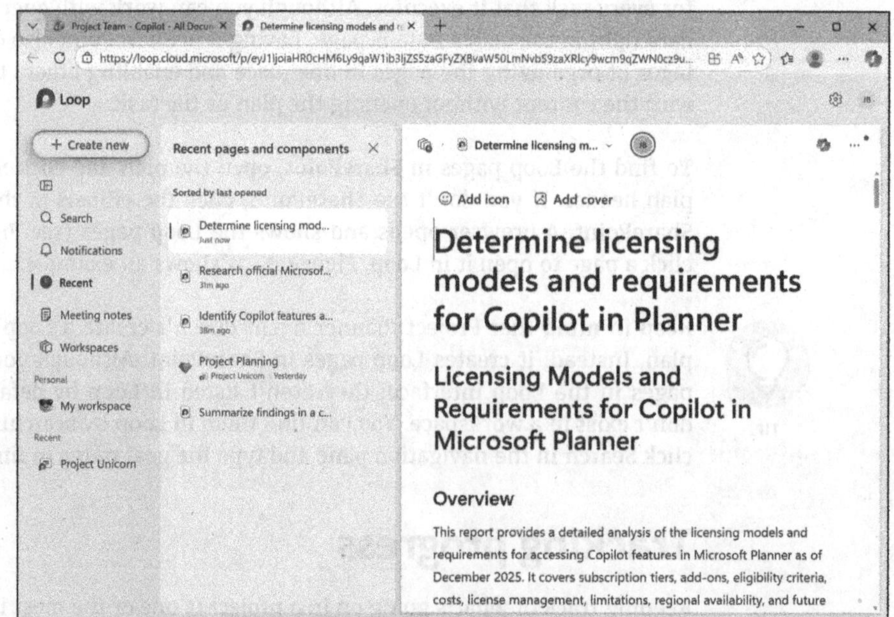

FIGURE 14-11: This Loop page contains the same information that the agent included in the corresponding task.

First, AI systems provide the best results when presented with the most possible data, so ensuring that your plan is fully fleshed out will give you the best results from Project Manager agent. Microsoft's documentation suggests that the plan should include at least ten tasks.

Follow these steps to generate a summary from Planner using Project Manager agent:

1. **Open a group Premium plan or a Plan with Project Manager plan that includes ten or more tasks.**

    In this example I use the Get Started with Project Planner plan I use earlier in this chapter.

2. **Click the Project Manager agent icon.**

3. **Click Understand, which inserts a prompt in the chat box (see Figure 14-12), and click the Send button (the blue right arrow in the lower-right corner).**

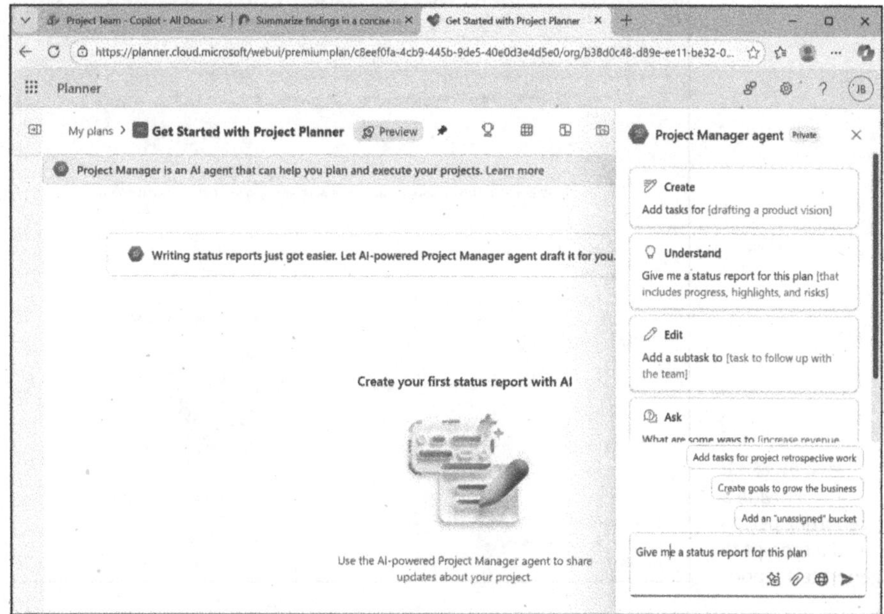

FIGURE 14-12: Click Understand to quickly generate a status report for your plan.

The agent responds with a lengthy status update that includes file references and links; a summary of email updates; insights from chats; and a summary of progress, highlights, and risks. The report, although it contains useful information, is somewhat limited given the current status of the plan. Only one task has been completed.

In addition, this is not a true report — it's just a summary response provided by Project Manager agent based on the prompt you provided. You can gather information about your plan in this way using a range of natural-language prompts. If the agent can't provide the information you're asking for, it will generally do a good job of explaining why and either provide suggestions or direct you to resources that can help you either find the information or refine your prompts.

Now, let's generate an actual report using Project Manager agent:

1. **Open Planner and open a group-shared Premium plan or a Plan with Project Manager Plan.**

2. **Click the Reports page.**

3. **Click the ellipsis to the right of the Project Manager agent message and click Get Started.**

   Planner shows the report prompt shown in Figure 14-13.

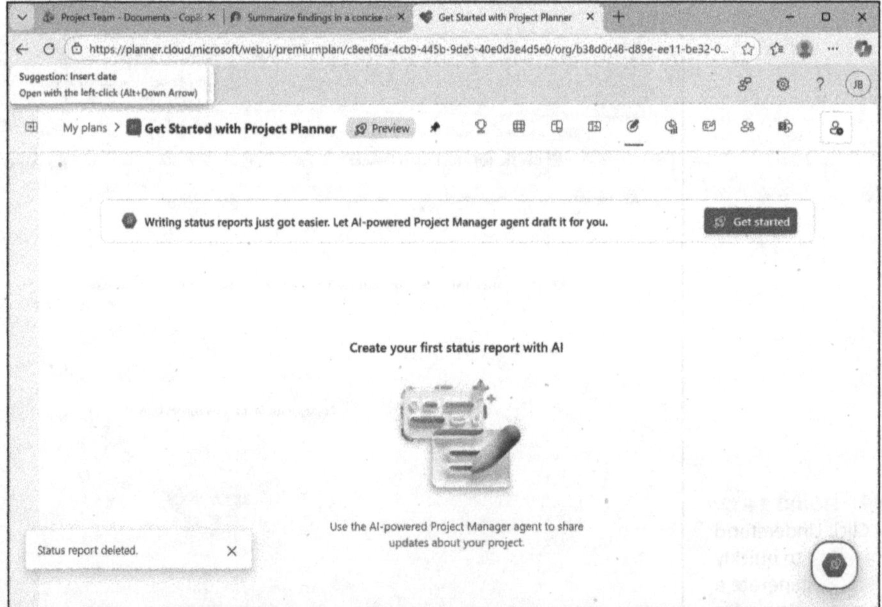

FIGURE 14-13: Provide a reporting period and a prompt to generate a report.

4. **Choose a reporting period from the drop-down list.**

5. **In the Additional Instructions field, type the prompt that defines the information you want to see.**

   In this example, I used the prompt, "Create a report based on task assignment and progress. Identify items that are on the critical path and what risks exist."

Figure 14-14 shows the resulting report in Planner. The report includes an Executive Summary, a suggested status (in this case, At Risk), a chart showing completion data, a chart of critical path timeline, highlights and achievements, challenges and delays, work in progress, and much more. Because the report is a Loop component, you can click the report and add your own notes or other information.

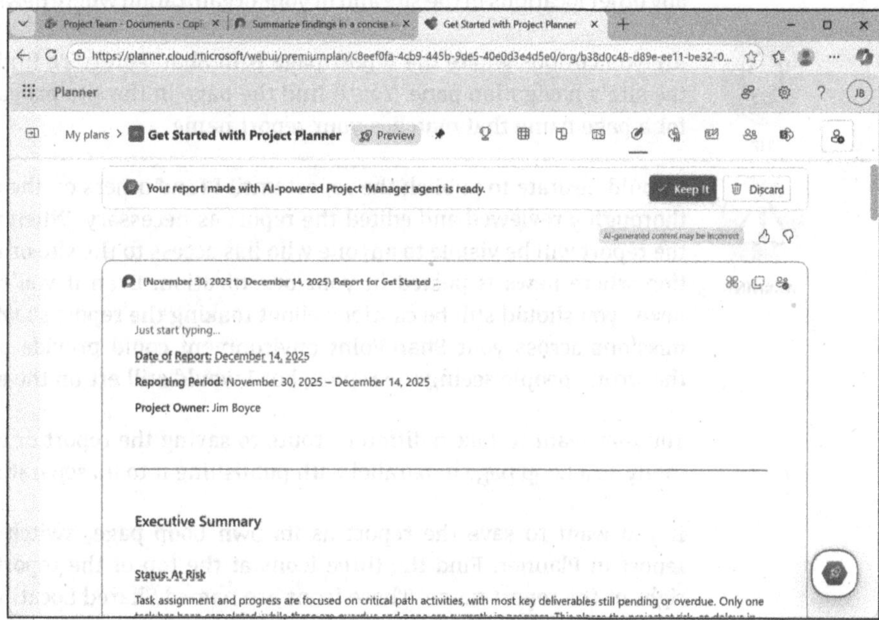

FIGURE 14-14:
The agent displays the resulting report in Planner.

You can review the report in Planner and decide if it provides the depth and type of information that you need. If you don't need to save the report or you want to generate a new one with different requirements, click Discard. Otherwise, click Keep It to save the report. When you do so, the agent gives you the option to share the report as a newsletter, choose a report status (such as Draft), or close the report (see Figure 14-15).

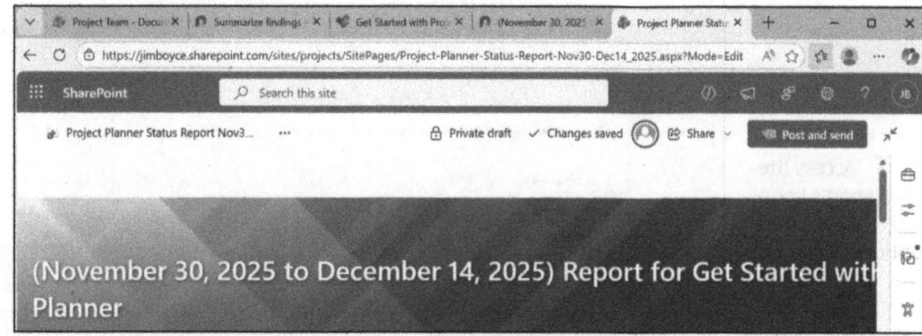

FIGURE 14-15:
You can set the report status and optionally share it as a newsletter.

If you choose the option to create the newsletter, the agent creates a Loop page in SharePoint with the contents of the report. The report is a private draft, but you can save it and then share it as needed. If you like the report, click the Keep It button. Make any updates or changes as needed, and if appropriate, have others on the project team add their updates to the report. Then use the Share menu to share it, or click Post and Send to post the report as a news item on the site's start page and any other locations in the site and in your organization where news items are posted.

TIP

Are you curious where the report page is in SharePoint? To find it, click Pages in the site's navigation pane. You'll find the page in the site pages library. Just look for a page name that matches your report name.

WARNING

I would hesitate to publish the report until I (and others on the project team) had thoroughly reviewed and edited the report as necessary. When you do publish it, the report will be visible to anyone who has access to the site or to any other location where news is posted in your organization. Even if you're reporting great news, you should still be cautious about making the report so widely visible. Permissions across your SharePoint environment could provide protection against the wrong people seeing the report, but I would still err on the side of caution.

You may want to take a different route to saving the report or even save it separately as a Loop page in parallel with publishing it to its separate SharePoint page.

If you want to save the report as its own Loop page, switch back over to the report in Planner. Find the three icons at the top of the report's header, to the right of the report name. These icons are named Shared Locations, Copy Component, and See Who Has Access. Click Shared Locations to view the options shown in Figure 14-16. Clicking the report's URL opens the report page in the Loop interface. If you want to create a Loop page for the report so you and others can find it more easily, click Add to Loop Workspace. You'll be prompted to choose an existing workspace or create a new one to store the Loop page.

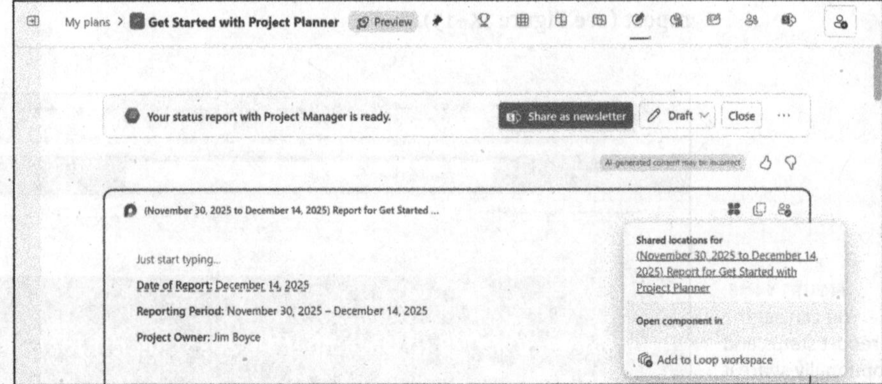

FIGURE 14-16: Access the report's Loop page and optionally store it in a Loop workspace.

Building good reports

The examples I explore in the preceding section are relatively simple, but they help to illustrate the ease with which you can create reports using Project Manager agent. Good reports depend on data that meets three requirements: accuracy, freshness, and depth. The more data the agent has to work with, the greater the insights and accuracy it can provide on the project's status. If the data is stale, the report won't reflect current state, which not only means that the report will reflect old data but that any recommendations or risk callouts will likely be inaccurate.

Meeting the data requirements I outline here is one aspect of generating great reports. The other, in this case, is ensuring that Project Manager agent has a clear understanding of the information you're looking for. Don't hesitate to create lengthy but clear and concise prompts, being very deliberate in the prompt about the information you need and how the agent should present it to you. Keep in mind that even if it takes you five minutes to craft a great report prompt, you're still saving hours of work compiling the information and drafting the report yourself.

Get in the habit of saving your carefully crafted AI prompts for future use and consider sharing them with colleagues. Deriving useful and accurate information from an AI is highly dependent on the clarity, accuracy, and detail in your prompts.

Chapter **15**

Reporting from Microsoft Planner

Whether you're tracking a small project for yourself or a large enterprise project, you need to be able to get good data out of your project planning tool and consume that data in an easily digestible way. This is important for everyone involved in the project and is even more important for rolling up status to your organization's leadership. This chapter focuses on visualizing data within Microsoft Planner through the application's built-in views, charts, and reports.

Visualizing Your Plan with Views

Planner provides reporting capability through Charts view and Reports view, both of which I cover later in this chapter. First, let's take a look at the things you can do within Planner's views to get a better handle on project status without using charts or generating reports. After all, you'll spend much more time working in

Planner than you will viewing charts and creating reports. When you get good at manipulating Planner's views, you'll likely need the reports less for yourself because you'll be able to view and understand the status of your projects without those charts and reports.

The first skill to grasp is filtering and grouping.

Using filters and grouping options

Planner's Grid, Board, and Timeline views all give you the ability to filter and group the tasks in the view to focus on specific items or visualize the data in different ways. Here are a few of the views you may want to see in Planner:

>> **Tasks assigned to specific people:** Work with the tasks assigned to you or check the status of tasks assigned to someone on your team (or to multiple team members).

>> **Tasks with a given priority or progress:** Check all high-priority tasks for status, find tasks not yet started, or review completed tasks.

>> **Late tasks:** View across all tasks the ones that are overdue so you can prioritize completion.

>> **Tasks due today:** Focus on all the tasks that are due on the current day to help you stay on track.

>> **Tasks aligned to a specific work category:** Understand the status of each work category in the plan, such as viewing the status of documentation, contracting, purchasing, and other specific work areas.

You can use filtering and grouping separately or in combination. For example, you could filter Grid view to show only tasks that are overdue. If you want to view the overdue tasks by work area (such as by bucket), then you would group the tasks as well as filter them.

Let's start with filtering. The following steps filter Grid view to show tasks with a finish date in the future using my Build a House project plan example:

1. **Open the plan you want to work with.**

2. **Open Grid view and click Filters (see Figure 15-1).**

   The Filter Tasks pane appears (see Figure 15-2).

3. **From the Finish Date drop-down list, choose Future.**

   The Grid view now shows only tasks with due dates after the current date.

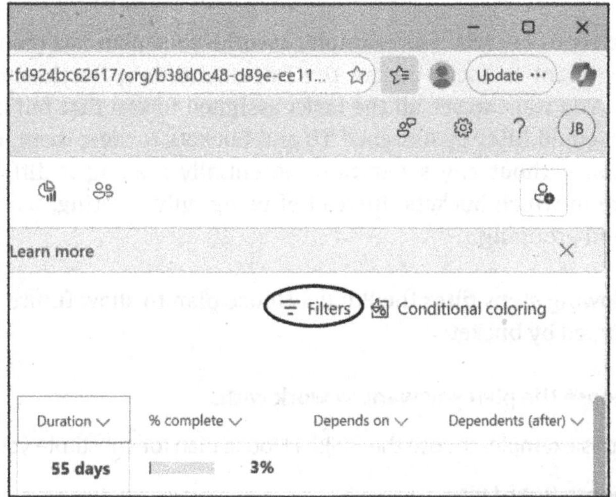

FIGURE 15-1: Click Filters in a view to choose filter options.

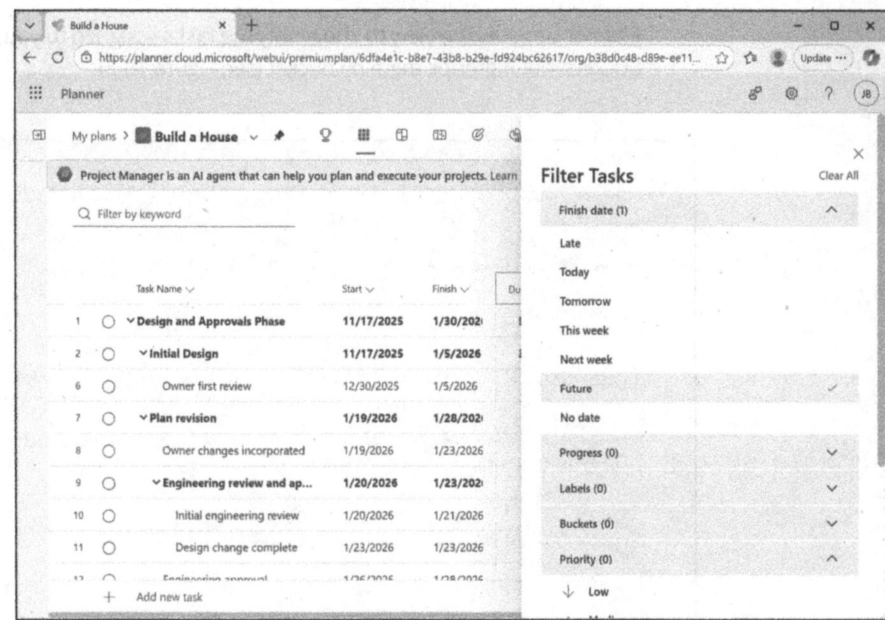

FIGURE 15-2: Choose filtering options from the Filter Tasks pane.

You can use multiple filter options to further focus your view. For example, after choosing the Future option for finish date, you might also choose Important from the Priority drop-down and your name from the Assigned To drop-down to view only the important tasks assigned to you with finish dates in the future.

Filtering is handy for limiting the amount of data shown so you can focus on a specific body of work. In addition to filtering, you can also group the data based on a variety of criteria. For example, assume your plan has many buckets but the ones you want to focus on are the Documentation, Purchasing, and Reporting buckets. You want to see all the tasks assigned to you that fall into these categories. You could filter by Assigned To and Buckets to view them, but they'll appear in the list without any separation, potentially making it difficult to see which tasks are in which buckets. Instead of using only filtering, use a combination of filters and grouping.

The following steps filter the Build a House plan to show future items assigned to me, grouped by bucket:

1. **Choose the plan you want to work with.**

   In this example, choose the Build a House plan (or substitute your own plan).

2. **Choose Board view.**

3. **Click Filters and choose Assigned To; then click your name.**

   Planner updates the view to show only the tasks assigned to you in the three buckets, but it groups them by bucket (see Figure 15-3).

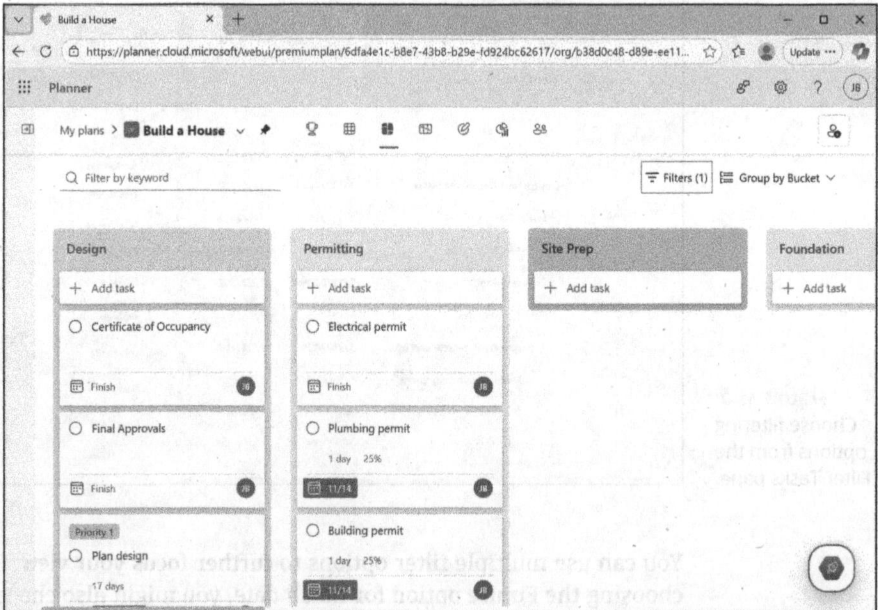

FIGURE 15-3: Board view filtered to show only tasks assigned to me.

In this example, I used Board view, which, by default, groups tasks by bucket. I didn't need to do anything other than filter the view, because it was already grouped by bucket. The Design and Permitting phases are mine. All the tasks in the Site Prep, Foundation, and other buckets are hidden because they're either unassigned or assigned to someone else. Trust me, you don't want me doing the foundation work. We don't want the house to fall down.

Now, let's group by a different property:

1. **In Board view, click the Group by Bucket drop-down list.**

2. **Choose Priority.**

I didn't change filtering, so Board view still shows only those tasks assigned to me. However, now it groups them by priority rather than by bucket, as shown in Figure 15-4.

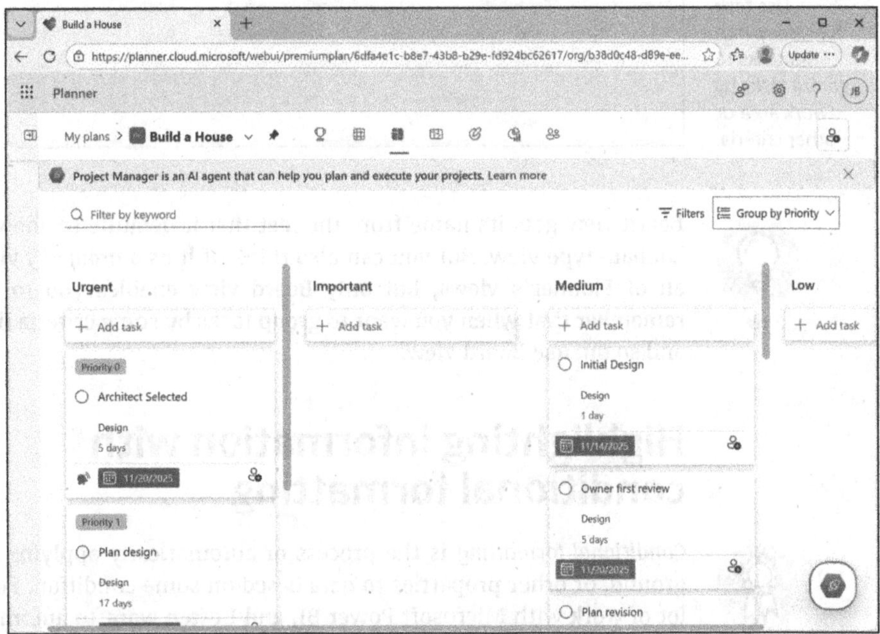

FIGURE 15-4: Use filtering and grouping to further refine a view.

PREMIUM

Grid view doesn't offer the ability to group. Instead, it shows tasks based on their task sequence. If you create subtasks, however, you'll see a sort of grouping, where Planner indents the subtasks under the higher-level tasks (see Figure 15-5). Using subtasks is a great way to organize your tasks into work area, categories, or other high-level criteria. You can then use buckets to further organize the tasks. Turn to Chapter 10 to learn how to promote tasks and create subtasks (which is a Premium feature).

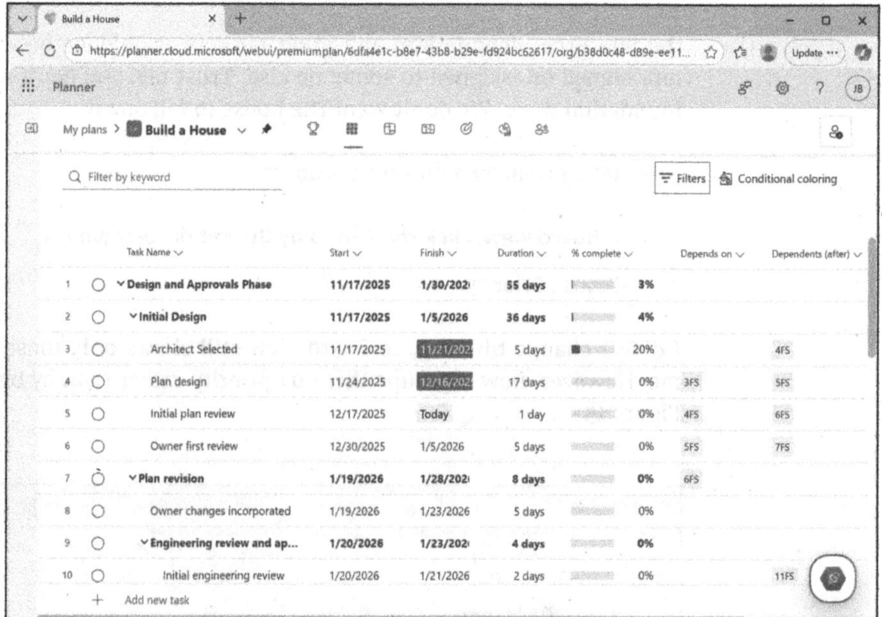

FIGURE 15-5:
Use task hierarchy with subtasks to organize tasks by work area or other criteria.

TIP

Board view gets its name from the fact that it defaults to showing tasks using a kanban-type view. But you can also think of it as a *group-by* view. You can filter all of Planner's views, but only Board view enables you to group tasks. Just remember that when you want to group tasks by some criteria like bucket, sprint, and so on, use Board view.

Highlighting information with conditional formatting

PREMIUM

Conditional formatting is the process of automatically applying font color, background, or other properties to data based on some condition. For example, I do a lot of work with Microsoft Power BI, and I often want to automatically apply cell background color to numeric values in a table based on their value, such as red for negative and green for positive. Or, maybe I'm showing month-over-month data, and I need to indicate when a value in the current month is higher or lower than the previous month. I could make the cell background of the current month green if the previous month was lower, indicating growth from last month to this month. Or maybe I would add an up arrow to the current month's cell to reflect the same thing. Whatever the case, these are all examples of conditional formatting. I specify a condition, and the application automatically applies formatting to data that satisfies that condition.

Planner offers the ability to use conditional formatting in Grid view. Grid view automatically colors the finish date in red if the task is overdue and yellow if it's due today. You can add your own conditional formatting for other conditions. For example, you may want to highlight tasks that have Urgent priority.

Conditional formatting works in Grid view, but not in other views. It's also a Premium feature — it doesn't work in Basic plans. In addition, you'll possibly experience issues with conditional formatting in Premium plans that are created from Planner's Premium templates or copied from a plan that uses a Premium template, where conditional formatting doesn't render at all. For now, if you have problems with conditional formatting, your only option is to create a Premium plan from scratch. Conditional formatting *should* then work in that plan, but there's no guarantee. Bottom line: Conditional formatting isn't fully baked in Planner as I write this.

Follow these steps to apply conditional formatting to the Grid view of a Premium plan:

1. **Open a Premium plan.**

2. **Open Grid view.**

3. **Click Conditional Formatting.**

   The Conditional Coloring dialog box appears (see Figure 15-6).

Conditional coloring

Priority | is equal to | Urgent

+ Add Color

Save   Cancel

FIGURE 15-6: Create formatting rules in the Conditional Coloring dialog box.

4. **From the Field drop-down menu, choose the field on which to base the condition.**

5. **From the Operator drop-down menu, choose the operator to apply to the condition (is less than, is equal to, and so on).**

6. **From the Value drop-down menu, choose the value to apply to the condition.**

7. **Click the color icon just to the left of the Delete icon and choose a color for the rule.**

8. **Click Add Color if you want to add another rule, or click Save to save the conditional format rule.**

When conditional formatting works, Planner highlights the background of the task name in Grid view using whichever color you selected for the rule.

Tracking Progress with Charts and Dashboards

You don't always need to create a report to understand what's going on in your plan. Using views, filtering, grouping, and conditional formatting as described in the previous sections gives you great flexibility in viewing your plans and tasks in Planner, and it can go a long way toward helping you quickly ascertain status and project health. Even so, the ability to analyze your plan with charts and reports is important as plan complexity grows. It's difficult to get a clear picture of a complex plan just by looking at Grid view, for example. Instead, you need to roll up complex data and render it visually or in report format to make it understandable and more easily consumed.

Charts view in Planner offers preconfigured visuals in a dashboard format. Basic plans include the following visuals in the dashboard:

>> **Status:** Shows a summary of the tasks by completion status

>> **Buckets:** Shows the number of tasks in each bucket along with their completion status

>> **Priority:** Shows tasks by priority, colored by completion status

>> **Members:** Shows the number of tasks assigned to each team member (and those that are unassigned), colored by completion status

Figure 15-7 shows an example of Charts view for a Basic plan.

A Premium plan offers similar Status and Bucket visuals and adds an Effort per Person visual to the dashboard. This visual shows the total effort of all tasks assigned to each person and is based on the Effort fields for each task. The visual is handy for identifying people who may be overburdened in the project.

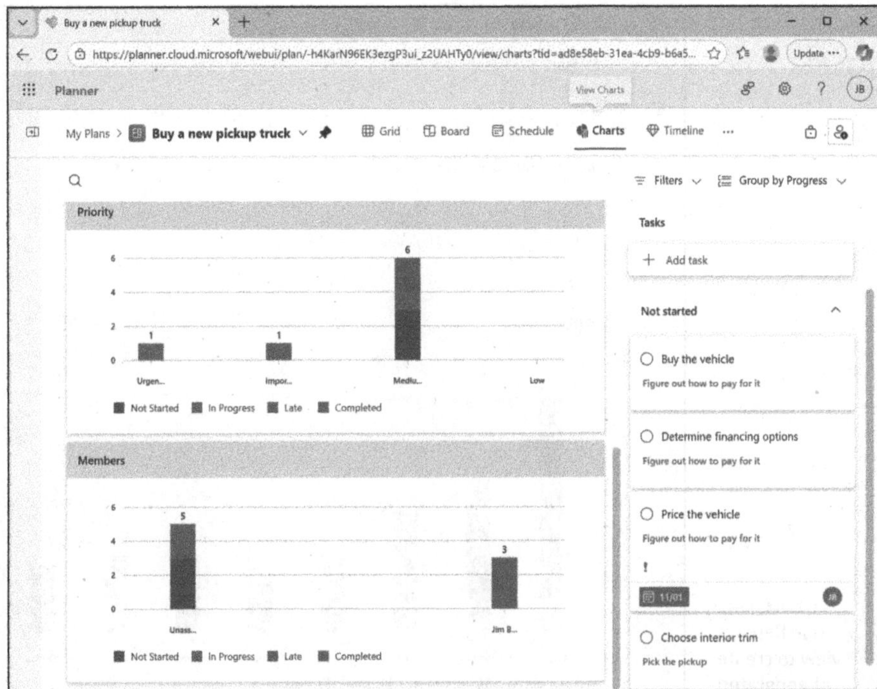

FIGURE 15-7: Use Charts view to visualize status by various criteria.

In my time at Microsoft, I used and built a lot of reports that tracked consumption, project status, metrics, and key performance results (KPRs). I have to say that the visuals offered by Charts view, while useful, are very rudimentary. Currently, you can't create custom visuals or customize the existing ones, other than apply filters to the Charts page. The visuals can certainly be useful, but the inherent reporting features offered by Charts view will quickly become insufficient to your needs. Fortunately, you have other options, starting with Microsoft 365 Copilot.

Using Copilot for Reporting

Chapter 14 explores Copilot, Microsoft's artificial intelligence (AI) suite that is built into many Microsoft 365 applications, including Planner. There, I briefly explain how to create a report using Copilot. This chapter explores the topic in more detail by providing best practices for generating quality reports.

Copilot's capabilities are surfaced in Planner Premium plans through Project Manager agent and Reports view (see Figure 15-8).

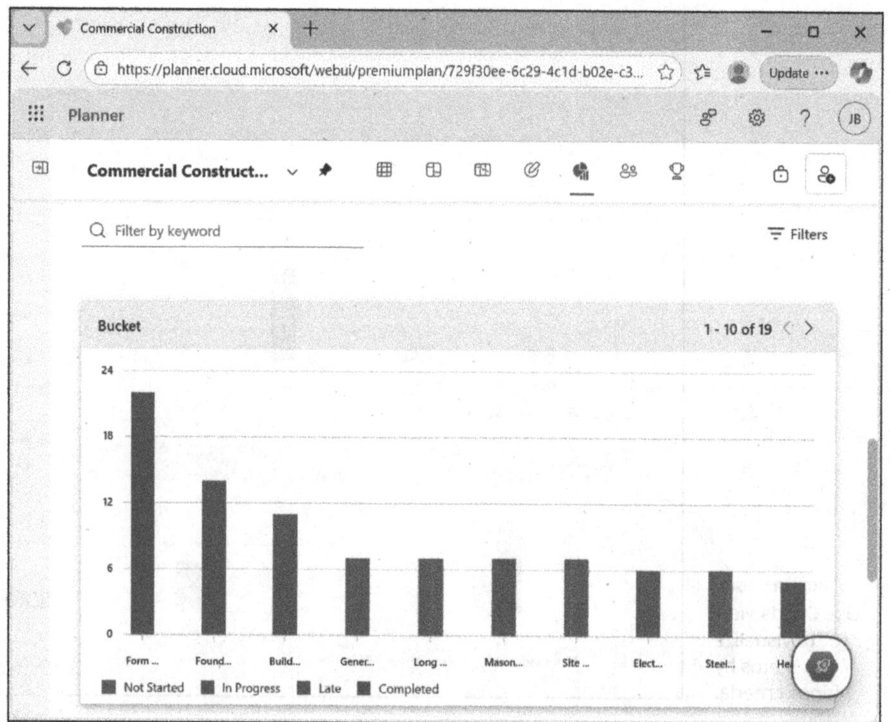

FIGURE 15-8: Use Reports view to create AI-generated reports.

Follow these steps to have Project Manager agent create a report for you:

1. **Open the plan for which you want to generate the report.**

2. **Open Reports view.**

3. **Click the ellipsis to the right of the text in Reports view and choose Get Started.**

   Planner shows the reporting options (see Figure 15-9).

4. **From the Reporting Period field, choose a time period for the report.**

5. **In the Additional Instructions field, type the prompt that tells Copilot what you want to see in your report.**

6. **Click Generate.**

7. **If you want to save the report, click Keep It.**

TIP

Planner saves the report to a Microsoft Loop page, which you can search for by name at https://loop.microsoft.com. You can also find the report in the Microsoft SharePoint site that is attached to the Microsoft 365 Group aligned to the plan.

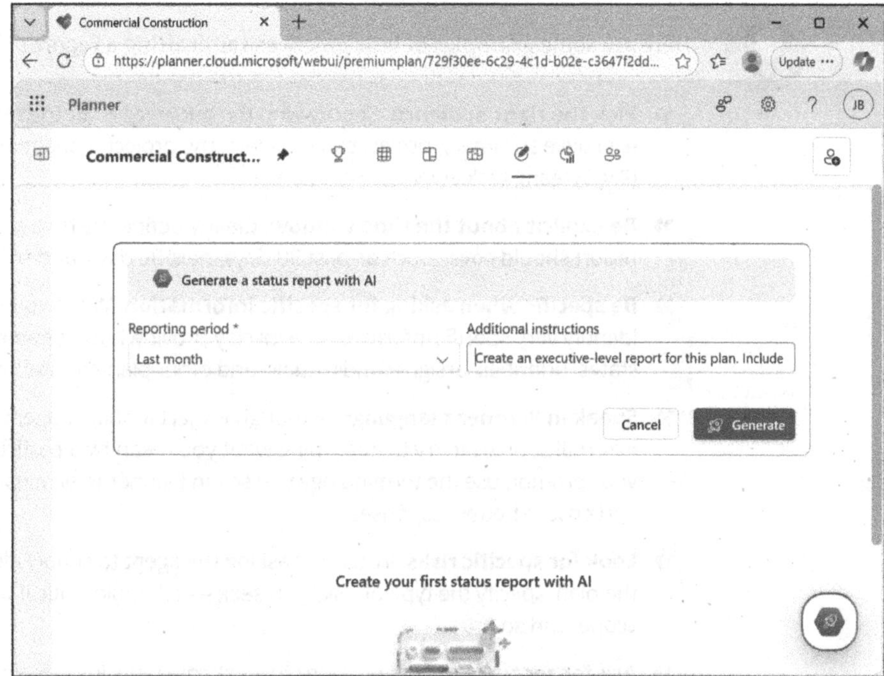

Project Manager agent will analyze the plan and create a report for you, but what you get in the report is almost entirely driven by the quality of the prompt that you give the agent in the Additional Instructions field. The more information you provide, the more detailed and accurate the report will be. For example, here are two prompts that create a status report, but the second will be much more detailed and useful:

Create a status report for this plan showing progress and potential risks.

Create an executive-level report for this plan. Include separate health indicators for schedule, workload, and risk. Highlight key accomplishments over the previous month. Identify milestones coming up in the next 30 days. Highlight items on the critical path that could delay project completion. Include a list of items that require leadership attention.

The second prompt will generate a much more useful plan, but one that is targeted at your executives. You can tailor your prompts to different audiences and scenarios, as well as different plan types. For example, if you want to reference specific buckets or other data in the plan, do so by name in your prompt.

Here are some examples of best practices for drafting a report prompt for Planner:

>> **Pick the right audience.** Specify who the audience is for the report whether executive summary, project stakeholders, the project management office (PMO) team, task owners, and so on.

>> **Be explicit about the time window.** Clearly define the time span that the report should cover, such as past 30 days, next 30 days, and so on.

>> **Be specific when asking for specific information.** When you need to identify very specific information, explicitly provide bucket names, progress states, priorities, assigned individuals, and other plan-specific information.

>> **Speak in Planner's language.** Although Project Manager agent can search external sources to try to determine what you mean by a particular phrase in your prompt, use the terminology you see in Planner to eliminate confusion and ensure better responses.

>> **Look for specific risks.** Instead of asking the agent to simply identify risks for the plan, specify the type of risks you seek — schedule, critical path, workload, scope, and so on.

>> **Ask for recommendations.** Don't just ask for status information. Leverage the power that Copilot offers by asking for recommendations in your report prompts.

>> **Stay focused.** Don't try to create a single report that covers every audience and report type. For example, create separate reports for executive summaries, workload analysis, risks, task status, and so on.

>> **Exclude and simplify.** Although you want to be explicit in your report prompts, explicitly exclude information that you *don't* need to see in the report.

>> **Tailor the prompt to the plan.** If you're reporting on a plan created from a template that has unique features (bucket names, labels, and so on) or one in which you've created custom fields, call out those fields, buckets, labels, and other elements explicitly in your report prompt.

>> **Specify what you want the report to look like.** Project Manager agent can create some great-looking reports, but unless you tell it what you want, the resulting report may not fit your needs. For example, if you want it to include a table of information, specify the column names and how you want the table formatted. Or tell it to use bullet points for action items or a list of risks.

Turn to Chapter 14 to explore how Project Manager agent leverages Loop components to create reports and how to save and work with these reports in SharePoint and Loop.

Extracting Data from Planner

The charting and reporting features in Planner — particularly those offered by Project Manager agent — can be very useful in helping you analyze project plans and generate reports. There will come a time, however, when you'll want to export data out of Planner for use in other applications or reporting tools.

Exporting a plan to Microsoft Excel is the most direct and easiest way to copy data out of Planner. You do so through Grid view:

1. **Open the plan that you want to export to Excel.**

2. **Click the drop-down menu beside the plan name in Planner's page header.**

3. **Choose Export to Excel.**

Planner exports your plan to an Excel file in your Downloads folder. Figure 15-10 shows an example of my Commercial Construction plan exported to Excel. The data includes some summary information like plan name, plan owner, dates, duration, and so on. Below that is a list of all tasks with name, assignments, duration, status, bucket, sprint, goal, and other information.

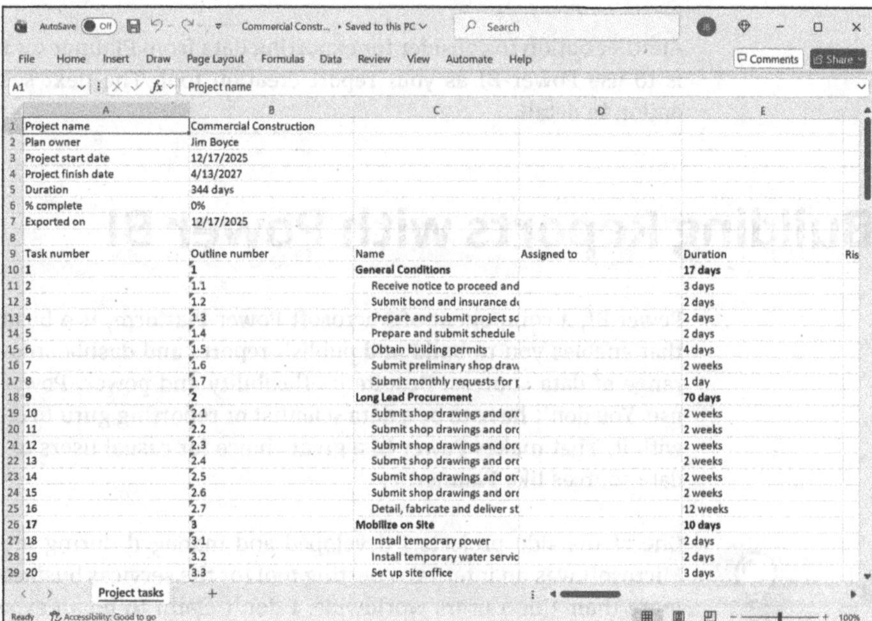

FIGURE 15-10: You can export a plan to Excel.

When you export data from Planner using this method, you get the entire plan. There is currently no method in Planner to filter and then export, so if you need only a subset of the data, one option is to export the data and then filter it in Excel or filter the data when you import it into your other application or reporting tool.

Exporting to Excel isn't the only option, however. You can use Microsoft Power Automate to build flows that export data, and you can tailor the flows to export specific data, like all tasks in a specific bucket, all tasks assigned to an individual, and so on. Exporting using Power Automate is more complex than exporting to Excel; it also requires some experience creating flows. Creating this type of flow is beyond the scope of this book, but I provide an introduction to Power Automate to help get you started in Chapter 13.

This is certainly a low-tech method, but you can export data from Planner from Grid view by copying and pasting. Select the data you want to export, press Ctrl+C, and then switch to your other application and paste the data.

The Microsoft Graph application programming interface (API) also offers a method for exporting data from Excel. It offers the most capabilities and control, but it's also the most technical and requires the most development experience of all the methods. For that reason I don't cover it here. You can find a detailed explanation with examples at `https://learn.microsoft.com/en-us/graph/planner-concept-overview`.

Another option to consider for exporting data from Planner for reporting purposes is to use Power BI as your report creation tool. The next section explores this option in detail.

Building Reports with Power BI

Power BI, a component of Microsoft Power Platform, is a business analytics tool that enables you to build and publish reports and dashboards from a very broad range of data sources. Despite its flexibility and power, Power BI is also easy to use. You don't have to be a data scientist or reporting guru to create useful reports with it. That makes Power BI a great choice for casual users to build reports from data sources like Planner.

One of the side projects I developed and managed during my last five years at Microsoft was an internal reporting tool for the services business that was used by more than 3,000 users worldwide. I don't claim to be an expert in reporting or Power BI, but even so, I was able to create a very useful tool that consolidated and simplified a wealth of data, thanks to how easy Power BI is to use. The key to success isn't necessarily your Power BI skills, but understanding the data, its relationships, and what you want to convey from that data. Don't be afraid to dig into Power BI even if you've never used a reporting tool like it before.

Getting an overview of Power BI

The Power BI Desktop application is what I primarily use to create reports and dashboards. You can also use the Power BI Service to build reports online using a web browser. Power BI Desktop provides a broader set of capabilities to ingesting data, building datasets, transforming and modeling the data, and building reports. The Power BI Service is where you publish, schedule report refreshes, and share reports; create dashboards; manage workspaces and apps; and do some light editing of published reports.

What's the difference between a Power BI report and a dashboard? A report is a multipage, interactive report that offers slicing, filtering, cross-filtering of visuals (where clicking an item in one visual can filter other visuals based on that selection), drill-down and drill-through, and tooltips and bookmarks. A dashboard brings together visuals from existing reports into a single-page canvas. Dashboards can incorporate visuals from multiple reports, making dashboards a great mechanism for creating executive summary reports and bringing together various aspects of a project.

In Power BI, *drill-down* refers to the ability to navigate through the various levels of hierarchy in a Power BI visual, digging deeper into the data. *Drill-through* lets you jump to another report page that's filtered to the selected value, like clicking a customer name in a report and navigating to a page that is filtered for that company.

What do you include in a Power BI report? Whatever you want as long as it's in the dataset! Great reports are designed with the end user in mind. For example, a Power BI report that is used by project managers will include visualizations that are different from an executive summary report that's designed for leadership. Certainly, some of the same information and visuals will appear on each, but each audience views the project in different ways and at different levels of detail. When you build your reports, seek extensive feedback from the report's audience before, during, and after development to make sure it's as useful as possible.

I strongly believe that the best reports are those that tightly engage the report audience during development. Over the years, I've seen way too many reports developed by data analysts with no experience in the audience members' jobs or roles. Often, the result is a boil-the-ocean approach that may show lots of interesting data but doesn't help the report users do their jobs effectively. Be very deliberate in building your reports with content and visuals that portray the data in the ways your audience expects to see it and, most important, fits seamlessly into the way they do their jobs.

Now, let's dig in and build a report using Power BI Desktop and an exported Excel file. If you don't already have Power BI Desktop installed, visit the Microsoft Store, search for Power BI Desktop, and install it. The application is free.

Building your first report from Excel

One of the easiest and quickest ways to build a report in Power BI is to import one or more Excel files. You can export data from Planner to Excel for Basic and Premium plans, making this approach a viable option for each plan type.

There used to be a Planner connector for Basic plans but it's no longer available. For that reason, exporting a Basic plan to Excel is the easiest way to obtain the data needed to build a report in Power BI. You can also use Power Automate to export data from Planner Basic plans to a SharePoint list or to an Excel file in Microsoft OneDrive. You can then import the data into Power BI from there.

This section assumes you've never used Power BI Desktop before to create a report. Don't worry! It's really easy to do.

Follow these steps to export your plan to Excel (I use my Build a House plan for this example):

1. **Open the plan you want to export.**

2. **Click the drop-down menu beside the plan name and choose Export to Excel.**

   Planner exports the data to your Downloads folder in an Excel file with the same name as the plan.

3. **Open your Downloads folder, find the file that Planner just created, and open it in Excel.**

   When you open the file, you'll see that it includes a summary section at the top with a table of tasks below it. Figure 15-11 shows an example of the Build a House plan exported to Excel.

   The "tables" in the Excel file are not true Excel tables yet, and I want them to be named tables so I can find them easily in Power BI. So, next, I need to create two tables for the data so that I can import those tables into Power BI.

4. **Insert a row at the top of the sheet, enter the text Item in cell A1, and enter the text Value in cell B1.**

   Item and Value will be the table column headers for the summary table.

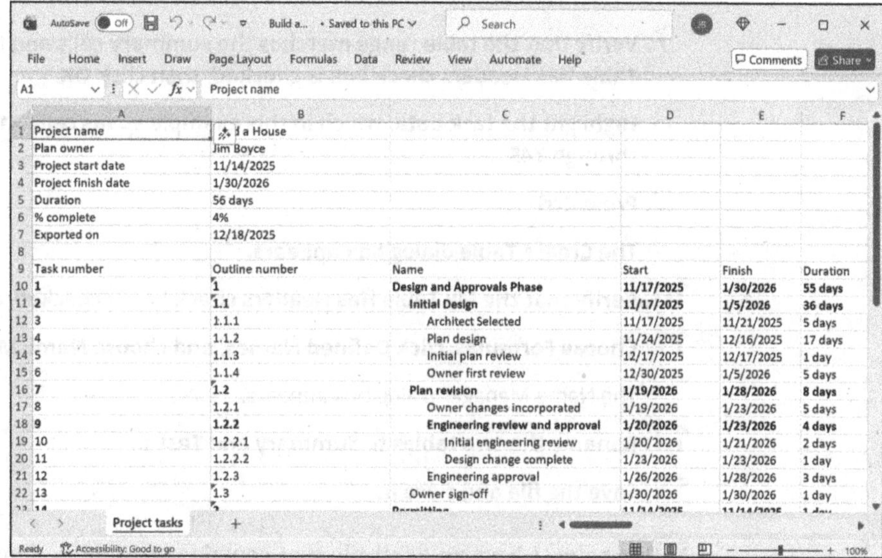

FIGURE 15-11: The Build a House plan exported to Excel.

5. **In Excel, with the Build a House Excel file open, click and drag to select all cells in the summary.**

In this example, the cell reference $A$1:$B$8 includes all cells in the summary data.

6. **Press Ctrl+T.**

The Create Table dialog box appears (see Figure 15-12).

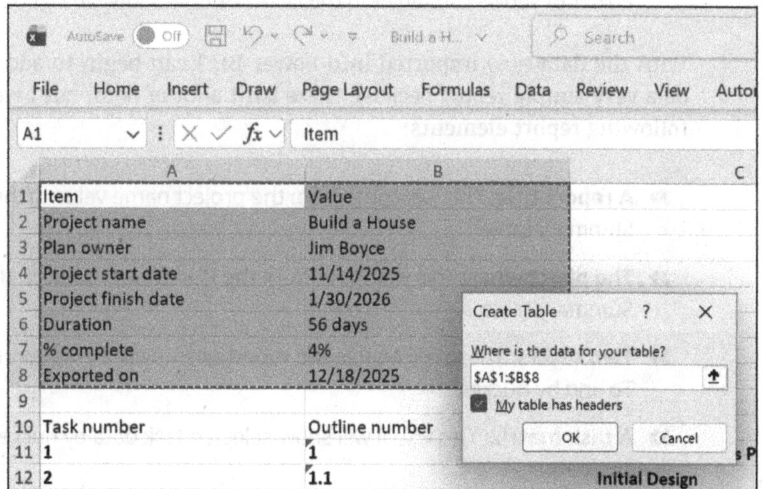

FIGURE 15-12: Create a table from the summary data.

7. Verify that the table range matches the summary cells and that the My Table Has Headers check box is checked; then click OK.

8. Highlight the task data, which in this example spans cells A10 through Y45.

9. Press Ctrl+T.

   The Create Table dialog box appears.

10. Verify that the My Table Has Headers check box is checked, and click OK.

11. Choose Formulas, click Defined Names, and choose Name Manager.

   The Name Manager dialog box appears.

12. Rename the two tables to Summary and Tasks.

13. Save the file and close it.

At this point, I have an Excel file that contains two tables, Summary and Tasks. Now my data is ready for importing into Power BI:

1. Open Power BI Desktop and click Excel Workbook in the list of data sources.

2. Choose the Build a House Excel file created previously and click Open.

3. In the Navigator window, select the check boxes for the Summary and Tasks tables (see Figure 15-13).

4. Click Load.

   Power BI shows two tables in the Data pane, as shown in Figure 15-14.

With the data now imported into Power BI, I can begin to add visuals. This will be a very simple report because there isn't a lot of data, but I want to include the following report elements:

>> **A report title:** This will come from the project name value in the Summary table.

>> **The plan owner:** This will come from the Plan Owner value in the Summary table.

>> **Two slicers:** These will enable the report user to slice the report by Assigned To and by Bucket.

>> **A task matrix:** This visual will show selected task data by bucket and name.

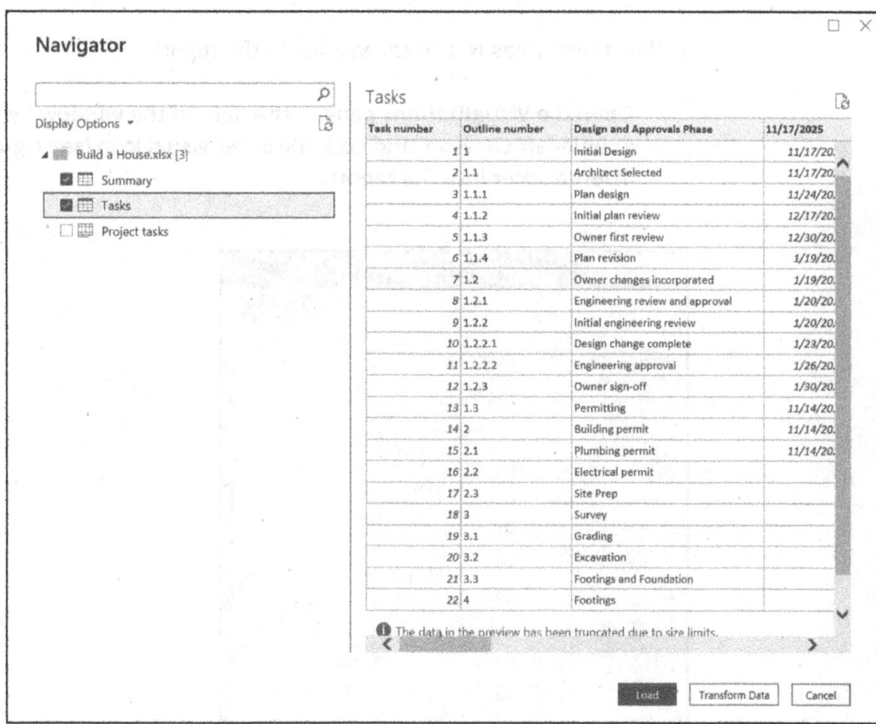

Task number	Outline number	Design and Approvals Phase	11/17/2025
1	1	Initial Design	11/17/20...
2	1.1	Architect Selected	11/17/20...
3	1.1.1	Plan design	11/24/20...
4	1.1.2	Initial plan review	12/17/20...
5	1.1.3	Owner first review	12/30/20...
6	1.1.4	Plan revision	1/19/20...
7	1.2	Owner changes incorporated	1/19/20...
8	1.2.1	Engineering review and approval	1/20/20...
9	1.2.2	Initial engineering review	1/20/20...
10	1.2.2.1	Design change complete	1/23/20...
11	1.2.2.2	Engineering approval	1/26/20...
12	1.2.3	Owner sign-off	1/30/20...
13	1.3	Permitting	11/14/20...
14	2	Building permit	11/14/20...
15	2.1	Plumbing permit	11/14/20...
16	2.2	Electrical permit	
17	2.3	Site Prep	
18	3	Survey	
19	3.1	Grading	
20	3.2	Excavation	
21	3.3	Footings and Foundation	
22	4	Footings	

FIGURE 15-13: Select the tables to import into Power BI.

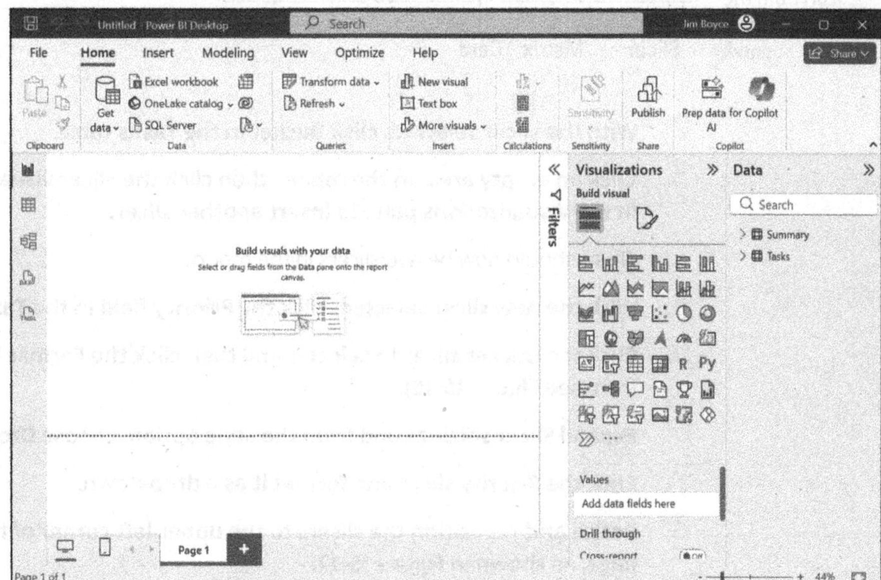

FIGURE 15-14: The two tables now appear in Power BI's Data pane.

Follow these steps to add the visuals to the report:

1. **Open the Visualizations pane at the right of the window by clicking the double-left chevron, and click the Slicer visual icon (see Figure 15-15) to insert a slicer into the report.**

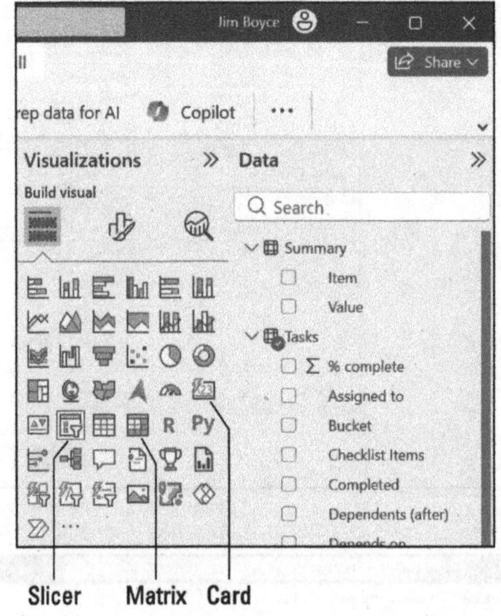

FIGURE 15-15: Choose the Slicer visual icon from the Visualizations pane.

2. **With the slicer selected, click Bucket in the Tasks table.**

3. **Click an empty area on the report; then click the Slicer visual icon again in the Visualizations pane to insert another slicer.**

   There should now be two slicers in the report.

4. **With the new slicer selected, click the Priority field in the Tasks table.**

5. **Click the Bucket slicer to select it and then click the Format Your Visual icon (see Figure 15-16).**

6. **Expand Slicer settings and from the Style option, choose Dropdown.**

7. **Click the Priority slicer and format it as a drop-down.**

8. **Resize and reposition the slicers to the upper-left corner of the report page, as shown in Figure 15-17.**

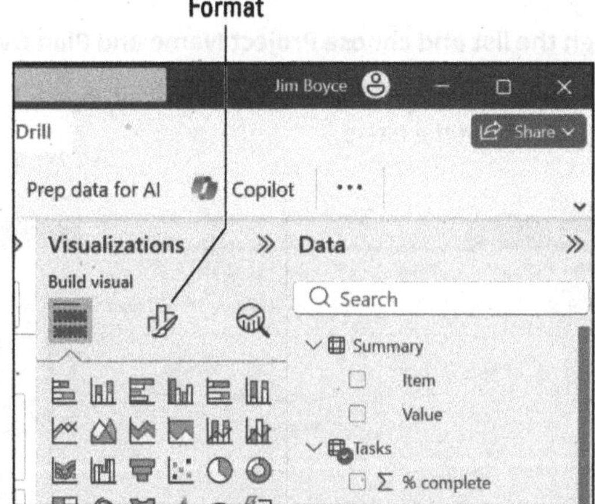

FIGURE 15-16:
Click Format Your
visual to open
the formatting
options.

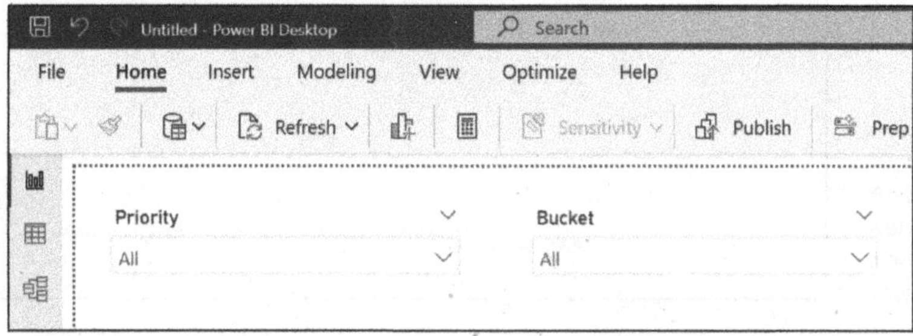

FIGURE 15-17:
Resize and
reposition
the slicers.

At this point you should have two slicers on the page that will enable you to slice the data by bucket and priority. For example, you may want to view only the Urgent items based on the Priority slicer, the items in the Design bucket, or some combination of the two.

Next, let's add a report title and owner in the upper-right corner of the report:

1. **Click in an empty part of the report; then in the Visualizations pane, find the Card visual icon and click it to add it to the report.**

2. **Drag the card to the upper-right corner of the report; then click Value in the Summary table.**

   The values of all the items in the Summary table appear in the card.

3. **Expand the Filters pane; then with the card visual still selected, drag the Item column from the Summary table to the Filters on This Visual section of the Filters pane.**

4. **Scroll through the list and choose Project Name and Plan Owner.**

   The card now shows the project name and plan owner. Figure 15-18 shows the resulting card visual and the filters.

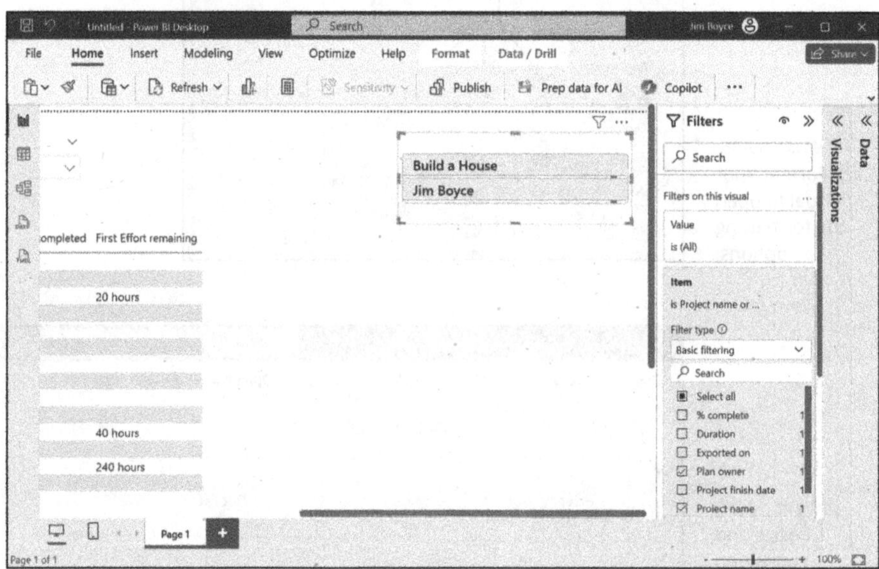

FIGURE 15-18: The card is now filtered to show only two items.

The last thing to add to this simple report is a matrix that shows the tasks by bucket and includes selected data from the columns in the Tasks table:

1. **Click in an empty space in the report and click the Matrix visual icon in the Visualizations pane.**

2. **Resize and reposition the matrix to use the majority of the remaining space on the report page.**

3. **With the matrix still selected, drag the Bucket column from the Tasks table in the Data pane to the Rows field in the Visualizations pane.**

4. **Drag the Priority, Duration, Effort, Effort Completed, and Effort Remaining columns from the Tasks table in the Data pane to the Values field in the Visualizations pane.**

   You'll need to drag these one at a time. Figure 15-19 shows the resulting matrix.

Bucket	First Priority	First Duration	First Effort	First Effort completed	First Effort remaining
⊞	Medium				
⊟ Design	Medium				
Architect Selected	Urgent	5 days	25 hours	5 hours	20 hours
Certificate of Occupancy	Medium				
Design change complete	Medium	1 day			
Engineering approval	Medium	3 days			
Final Approvals	Medium				
Initial engineering review	Medium	2 days			
Initial plan review	Urgent	1 day			
Owner changes incorporated	Medium	5 days			
Owner first review	Medium	5 days	40 hours	0 hours	40 hours
Owner sign-off	Medium	1 day			
Plan design	Urgent	17 days	240 hours	0 hours	240 hours
⊞ Foundation	Medium				
⊞ Framing	Medium				
⊞ Permitting	Medium				
Total	Medium				

FIGURE 15-19:
The matrix shows selected fields from the Tasks table.

That's it for this report. I confess that this is a very ugly report and doesn't reflect the level of data you'll need when you start creating your own Planner reports using Power BI. However, here are the key concepts that this exercise illustrates regarding Power BI reports:

>> It's easy to create a report from an Excel file using Power BI, even with no Power BI experience.

>> You can apply filters to visuals to limit the data that they show. (The same is true for pages — you can apply filters at the page and report levels to control what data appears in the report.)

>> You can quickly filter a report page by choosing one or more items from the slicers.

>> Your report is only as good as the data it visualizes. Several of the tasks in the matrix on this report are missing effort information because the underlying data is incomplete. More data equals better, more accurate reports.

Integrating multiple data sources

In some cases, you'll need to import data from multiple sources. For example, if you're using Excel as the data source for your report and you want to include multiple plans, you'll need to add multiple Excel files to the report, one for each report. Or you'll have to combine the individual Excel files into one file so you can ingest the data from a single file.

If you decide to use multiple files, just import each file separately. You'll end up with tables from each file, which also means you'll probably need to rename the tables after you import them so you know what's what. To do so, click the ellipsis to the right of the table name in the Data pane, click Rename, and enter a new name.

You're not limited to just adding multiple Excel files. Power BI can combine data from many different sources. In this case, you might add Excel files, SharePoint lists, or other data to a report. Depending on the data and how you structure the report, you may not need to do anything to the data other than import the various datasets. However, in some cases, you'll need to establish relationships between tables.

Consider the example I describe earlier, where you have a single master file with summary information about all the plans and a file that compiles all the tasks for all the plans, one task per row. Create a column in the master file with a field named PlanID; then assign each plan a unique ID. It can be as simple as an integer, but each must be unique. Then add a column in the plan task file called PlanID and for each row assign the unique value you defined for the task's plan in the master file. The PlanID column in the tasks table now identifies the plan in which the task resides.

Next, import both files into Power BI as described in the previous section. Add the relationships using the following method:

1. **Open Power BI and import both files.**

2. **Click Model View in the Power BI Navigation pane (see Figure 15-20).**

Model View icon

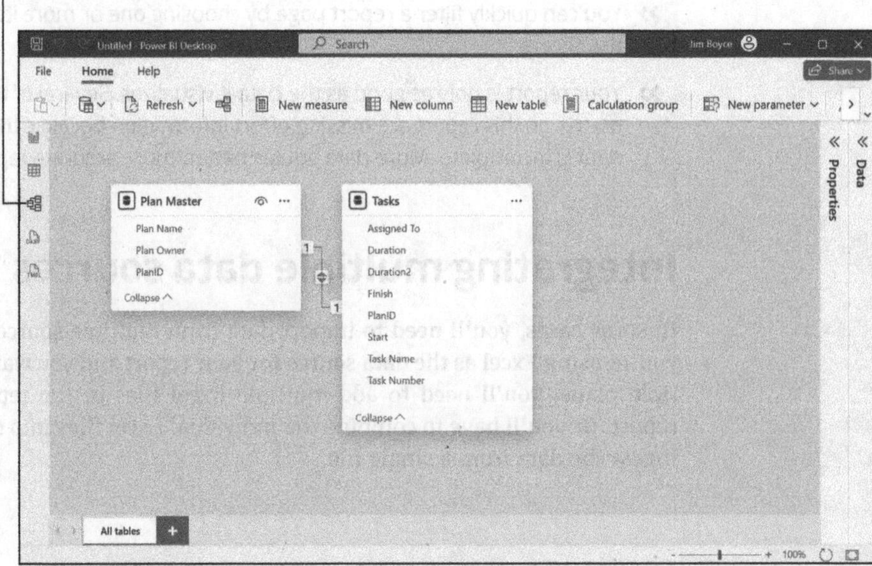

3. **If Power BI does not automatically detect the relationship based on the
column names in both files, click and drag PlanID from the Tasks table to
the PlanID column in the Plan Master table.**

In this example, Power BI automatically created a relationship for me because the
PlanID columns existed in both tables. Otherwise, creating the relationship is as
easy as dragging the PlanID column from one table to the other. Now, you can use
the Plan Master table to slice and filter the data in the Tasks table because a relationship now exists between the two tables.

I touch on relationships in this chapter to give you a very basic overview and get
you started down the right path. But it's a relatively complex topic that you'll need
to research further to go very far down the road. What's more, there are different
types of relationships. Have a look at `https://learn.microsoft.com/en-us/`
`power-bi/transform-model/desktop-create-and-manage-relationships` for
more details.

Ingesting data from the Dataverse

If you're working with Premium plans, you can still export data to Excel and
import the Excel files into Power BI, but there's another option. You can connect
directly to the plan data stored in Microsoft Dataverse.

Dataverse is Microsoft's cloud-based data platform used to store and manage data
across many of Microsoft's cloud applications including Planner Premium plans.
You can connect to Dataverse from Power BI to access plan data and build reports
based on that data. Here are some key licensing considerations for using Power BI
in combination with Dataverse-backed data:

» No special license is required to connect to Dataverse-backed data from
Power BI to create reports. This enables you to experiment with creating
reports without any additional investment.

» You must have either a Power BI Pro or Power BI Premium Per User (PPU)
license to publish a report to the Power BI service to share with others.

» Building reports with advanced features such as large models and AI features
requires a Power BI PPU license.

» People viewing a report in a shared non-Premium workspace need either a
Power BI Pro or Power BI PPU license to view the reports.

» For workspaces in PPU mode, each viewer much have a Power BI PPU license.

> If the report is published to a Premium-capacity workspace, all viewers can use a free Power BI license to view the reports. The report publisher must still have a Pro or PPU license to publish to a Premium workspace.

Here's how to connect to Dataverse in Power BI Desktop to use your Premium plan data:

1. **Open Power BI and start a blank report.**

2. **On the Home tab click Get Data, and then choose Dataverse (see Figure 15-21).**

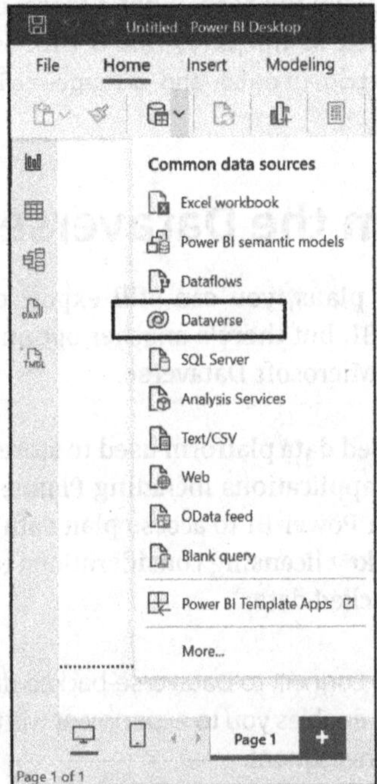

FIGURE 15-21:
Connect to Dataverse to access Premium plan data.

3. **Expand the table list in the Navigator and search for the tables that start with msdyn_ in the table name.**

4. **Choose each of the tables that you want to include, based on the data needs of the report.**

5. **Click Load to load the data.**

    After a short period time, the data will be imported into Power BI and you'll see the resulting tables in the Data pane.

There are several things to consider here. For example, which tables should you use? You don't want to simply bring in all the tables that start with msdyn_ into your model. There currently is no consolidated single-source reference for the Planner Dataverse tables, but Table 15-1 lists the ones you should investigate first.

Planner Dataverse Tables

Entity Name	Table
Project	msdyn_project
Project Task	msdyn_projecttask
Project Task Dependency	msdyn_prolejcttaskdependency
Resource Assignment	msdyn_resourceassignment
Project Bucket	msdyn_projectbucket
Project Team Member	msdyn_projectteammember
Project Checklist	msdyn_projectchecklist
Project Label	msdyn_projectlabel
Project Task to Label	msdyn_projecttasktolabel
Project Sprint	msdyn_projectsprint

TIP

To explore further, simply type **msdyn_project** in the Power BI Navigator to view all tables whose names start with msdyn_project.

When you import the data from Dataverse, the relationships between tables will be replicated in Power BI so you won't need to build your own relationships. For example, try this:

1. **Import msdyn_project and msdyn_projecttask to a new Power BI report.**

2. **Add a table visual.**

3. **Add the Name and Owner columns from msdyn_project to the table visual.**

4. **Add the Project Task Name, Owner, and Project Bucket columns from the msdyn_projecttask table to the same table visual.**

The visual now includes columns from both tables in a table format, and you didn't have to build any relationships to make it work because they already exist. Next try adding a slicer using the Name column from msdyn_project, which would list all the plan names. Click a plan name in the slicer and the table visual filters to show only the tasks associated with that plan.

For more information on Dataverse and how to work with it, see `https://learn.microsoft.com/en-us/power-apps/maker/data-platform/data-platform-intro`.

Chapter 16

Managing Microsoft Planner in the Enterprise

The majority of this book is structured for Microsoft Planner users, but there are several topics that Microsoft 365 administrators should know to effectively and securely enable, manage, and report on Planner usage in their organizations. This chapter explores key administrative activities for Planner to ensure effective implementation, strong security, good governance, and usage visibility.

Evaluating Planner

I've been writing books and articles on computers and technology for almost 40 years, and I can tell you with certainty that most people don't read computer books cover to cover. They pick out the bits they need at the moment and move on. If you're reading this chapter, you probably fall into one of two camps: Either you're a Microsoft 365 administrator who mainly needs to manage Planner and doesn't use it that much, or you're a power user who needs to manage it in your own *tenant* (your organization's unique space in Microsoft's cloud environment). Or maybe you use Planner or will be using Planner for your IT projects as well as managing it. Whatever the case, this chapter will help you understand from an administrator's perspective how Planner fits into your organization, how to deploy it, and how to manage it.

The chapters in Part 1 of this book provide a sound overview of what Planner is and how to get started using it. Those chapters are geared more toward the end user, but they can also give an administrator a good overview of Planner. In this section, I look at Planner from a slightly different perspective — that of a Microsoft 365 administrator or technology director who needs to decide how Planner can fit into their organization; explain it to leadership to get buy-in or justify costs; and understand the key points to ensuring that Planner is deployed economically, effectively, and securely.

Contrasting Planner and Project

As Chapter 2 describes, there are two "flavors" of Planner and four subscription types. The Planner flavors are Basic and Premium. Planner Basic is included with many Microsoft 365 subscriptions at no additional cost and is geared toward personal planning and light team project planning. Planner Plan 1, Planner and Project Plan 3, and Planner and Project Plan 5 are the Premium versions and require a specific subscription. The Premium versions add varying levels of true project-management capabilities, each building on the features in the next lower version. Planner and Project Plan 3 builds on Planner Plan 1, and Planner and Project Plan 5 builds on Planner and Project Plan 3.

Planner Basic offers simple task-management capabilities for individuals and teams. It offers a Grid view, Board view, and Schedule view. It includes the ability to assign tasks to others when linked to a Microsoft 365 group and to collaborate in other ways. In a larger organization, Planner Basic can be an effective solution for lightweight project-planning needs without additional cost. For example, Planner Basic is included in Microsoft 365 Business Standard, Business Premium, E1, E3, E5, A3, and A5 licenses. It's also available in Government Community Cloud (GCC) plans.

By contrast, the three Premium versions require a separate license, which adds cost for each user who needs those Premium features. The Premium features include support for timelines, task dependencies, custom fields, goals, backlogs and sprints, Premium templates, expanded reporting capabilities, Microsoft 365 Copilot integration, task history, road maps, baselines and critical paths, enterprise resource management, and enterprise portfolio management. Which features you have depends on which Premium subscription you're assigned.

Microsoft Project for the Web was retired by Microsoft in August 2025 and replaced by Planner Premium subscriptions. The Project and Microsoft 365 Roadmap apps in Teams were also retired. Planner integrates the features that were previously available in these offerings. Effective October 1, 2025, Microsoft no longer offers Project Online, and has set September 30, 2026, as the official retirement date for Project Online. What Microsoft is *not* retiring is Project Desktop or Project Server Subscription Edition.

Finding the data

Planner basic stores the following items in the Planner service, hosted in Microsoft Azure:

>> Tasks

>> Buckets

>> Assignments

>> Progress and status

>> Due dates

>> Labels

>> Checklists

>> Task-level metadata

For Basic plans that are tied to a Microsoft 365 group, Planner stores attachments in the group's Microsoft SharePoint site. Basic plans not tied to a group can only insert links to files or URLs, so there is no file storage support for these plans. Planner leverages Microsoft Exchange Online to store task comments in group conversations and delivers notifications through Exchange and Microsoft Outlook. When a task is assigned to a user, Planner synchronizes a copy of the task to the Microsoft To Do app. Microsoft Entra ID manages permissions for all actions and features.

Premium plans are different in that the Planner data such as plans, tasks, dependencies, milestones, and other data are stored in Microsoft Dataverse. Premium plans must be tied to a Microsoft 365 group, and Premium plans use the same collaboration features as group-enabled Basic plans for storing files (SharePoint), comments (Exchange Online), and notifications (Outlook and Exchange). So, the primary difference between the two in terms of data storage is that Basic plans use the Planner service in Azure and Premium plans use Dataverse.

Integrating Planner with other Microsoft 365 applications and services

As a Microsoft 365 application, Planner integrates with several other Microsoft 365 applications and services. As I describe earlier, for example, plans tied to a Microsoft 365 group store files in SharePoint. But Planner integrates with several other applications. Planner users can use the Planner web app to work with plans, or they can use the Planner app in Teams. Users are not limited to working with Premium plans in Teams — they can work with both Basic and Premium plans.

In Chapters 14 and 15, I explore the integration of Planner and Microsoft Loop. Project Manager agent, which is the Copilot agent for Planner, can be assigned and execute tasks. The data that the agent creates for the task is stored as a Loop page, and users can access the Loop page within the Planner web app or Teams, from `https://loop.microsoft.com`, or from the SharePoint library where the Loop objects are stored. By default, the SharePoint folder is `\Documents\ Microsoft Planner\[Plan Name]\Copilot`, where `[Plan Name]` is the plan name assigned in Planner. Reports generated by Copilot are also saved as Loop components and searchable at `https://loop.microsoft.com` based on the report name.

Other integrations include To Do, Microsoft Power Automate, and Microsoft Power BI. As I mention earlier, assigned tasks are automatically synced to the To Do app so users can work with their assigned tasks in To Do or Planner. You can automate a wide range of actions in Planner with Power Automate (see Chapter 13 for a detailed discussion). As I explore in detail in Chapter 15, you can use Power BI to build reports using data exported directly from Planner to Excel or data exported to Excel or SharePoint using Power Automate, or you can connect to data directly from Power BI to Dataverse for import or direct query report building.

Leveraging Purview

Microsoft Purview also offers some possibilities for integration with Planner. Purview doesn't manage Planner directly but rather manages the underlying services that Planner uses. For Planner Basic, this means using Purview to manage Microsoft 365 Groups. For Basic plan data, Purview can retain task comments via Exchange Online, attachments via SharePoint Online, group membership and metadata through Microsoft 365 Groups, and emails and notifications via Exchange. To control retention for Planner Basic plans, apply retention policies to Microsoft 365 Groups, SharePoint sites, and Exchange group mailboxes. However, you can't retain Planner task objects independent of the Microsoft 365 group.

eDiscovery with Planner Basic plans will surface task comments and group comments from Exchange and attached documents from SharePoint. Planner tasks are not searchable items in eDiscovery for Basic plans, which means that task titles, descriptions, due dates, assignments, and progress metadata are not discoverable. Following this same theme, data loss prevention (DLP) can enforce policies on files in SharePoint and emails and comments through Exchange.

Premium plans offer much deeper integration with Purview because they store their data in Dataverse. This means you can enforce policies to retain or delete tasks, dependencies, project metadata, and custom fields for Planner Premium plans. Likewise, eDiscovery is better with Premium plans, enabling you to discover task titles, descriptions, assignments, dates, status fields, and custom fields.

DLP is also much expanded with premium plans. Through Dataverse DLP policies you can:

>> Prevent data movement from Planner to non-approved connectors.

>> Prevent data movement between environments.

>> Block task data from being exported to personal apps.

>> Restrict Power Automate flows from accessing plan data.

>> Enforce boundaries at the tenant and environment levels.

Auditing and insider risk management are also possible with Planner Premium plans, enabling you to log items such as plan creation, task creation and updates, assignment changes, and item deletions. Insider risk management is supported through SharePoint, Exchange, and Dataverse signaling.

Depending on your organization and industry compliance requirements, the inability to provide all but very basic inspection, eDiscovery, data loss prevention, and auditing over Planner Basic plans could be the very reason you need to move to a Premium-subscription-only approach.

Deploying Planner

Like other Microsoft 365 applications, you don't really deploy Planner in the classic sense because there is nothing to install or configure specifically for Planner. Instead, deploying Planner in your organization means ensuring that the underlying Microsoft 365 services are enabled and users are granted the appropriate licenses that light up Planner for them. These services and licenses include:

>> Microsoft 365 licensing

>> Microsoft 365 Groups and group creation (to enable plan creation)

- >> SharePoint Online (file attachments)

- >> Exchange Online (comments, notifications, and group mailboxes)

- >> Entra ID

- >> Dataverse (Premium plan data storage)

- >> Power Platform (optional, for Power Automate and Power BI integration)

- >> Copilot (optional, for Project Manager agent features)

- >> Loop (for storing Copilot-generated task and report data)

- >> Planner Premium licenses (to unlock subscription-based Premium features)

With the assumption that you have a fully functioning Microsoft 365 environment, implementing Planner really becomes all about managing licenses. However, you can use the following as a general guideline:

- >> Ensure the appropriate underlying Microsoft 365 services are enabled and properly secured.

- >> Restrict Microsoft 365 group creation to those users who need to create plans (and those who need to create groups for other reasons other than Planner).

- >> Configure Purview policies as appropriate.

- >> Enable Planner through licensing to those who need it.

- >> Roll out to your users through the Planner web app, through Teams, or both.

Managing Access to Planner

Unless you're willing to allow anyone in your organization to use Planner (and have licenses to accommodate that strategy), controlling access is probably one of your concerns as an administrator. There are a handful of strategies, with the most common being controlling licensing. If users aren't assigned a Planner Premium subscription, they won't be able to use Planner Premium features. However, if they have Planner Basic through their base Microsoft 365 subscription, they'll still be able to create and use Basic plans. The most direct method for disabling Planner Basic for those users who have a Microsoft 365 license that includes it is to disable the Planner license in the admin center. Note that you can disable Planner Basic while still enabling Planner Premium for those who have the appropriate license. Figure 16-1 shows an example where the user's Basic license is disabled but the Premium license is enabled.

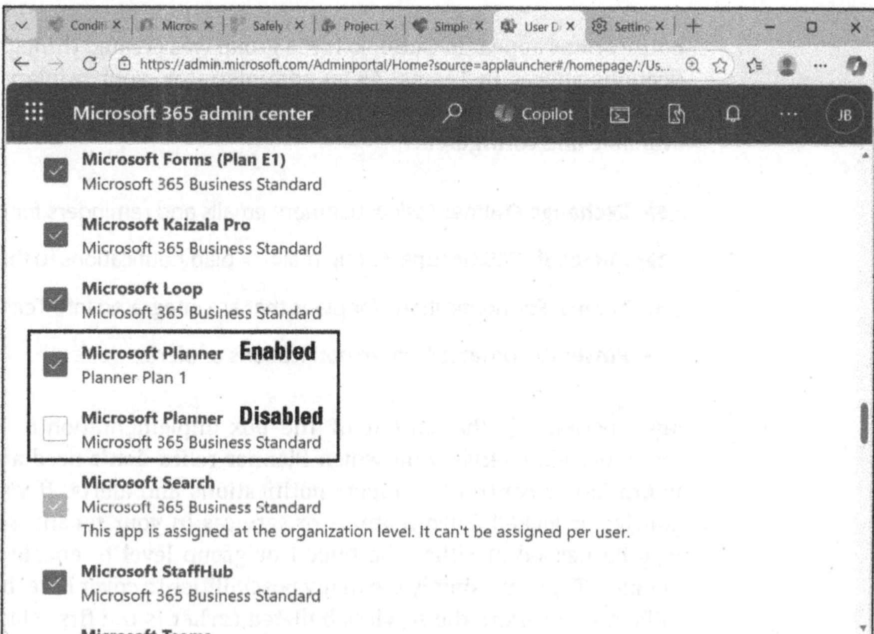

FIGURE 16-1:
You can disable Planner Basic without disabling Planner Premium.

In some scenarios, you may want to give external users access to Planner. For example, maybe you have a vendor that is helping with documentation, procurement, or other activities. External users can't create plans, but they can access and work with plans that are shared with them.

TIP

Here are tips for enabling external user access to Planner:

>> Invite the user as a guest in your Microsoft 365 tenant.

>> Use a group for the plan that allows guest participation, and add the user to the group.

>> To enable Premium features for the user, assign them a Planner Premium license.

>> Ensure security through conditional access and other Microsoft 365 Groups policies.

Managing Notifications and Alerts

Planner sends user notifications when tasks are assigned, tasks are due soon, task status changes, and comments are added. Where these notifications appear depends on how the plan's group was created. If it was created from Outlook, the notifications appear in Outlook. If the group was created through Teams, the notifications appear in Teams. As an administrator, your primary task for enabling and managing notifications and alerts is to ensure that the underlying services are available and configured:

>> **Exchange Online:** Task assignment emails and reminders for due dates.

>> **Microsoft 365 Groups:** Group mailbox plan notifications to the group.

>> **Teams:** For notifications for plans that are integrated into Teams.

>> **Power Automate:** Custom notifications or alerts.

My experience is that an out-of-the-box implementation of Microsoft 365 and the underlying services on which Planner relies don't need any additional configuration or control to support notifications and alerts. If you've implemented policies or locked down features or services in your tenant, some configuration may be needed at either the tenant or group level to enable these features for Planner. There are simply too many possibilities to cover here, but ensuring access to Planner users to the services bulleted earlier is the first place to look if you're having issues with notifications and alerts.

See Chapter 8 to learn more about how comments and notifications work from a user perspective.

Employing a Backup Strategy

Something, sometime, is going to break. That's not a threat or a promise; it's a statement of fact. That's why it's important that you have a backup and recovery strategy for everything, including Planner. That strategy varies depending on plan type.

Basic plan data is stored primarily in the Planner service in Azure, which has its own redundancy built in. There is no native task-level restore for Planner tasks, nor is there a capability to restore deleted plans independently. However, Microsoft does back up Microsoft 365 Groups data through service continuity and SharePoint documents are covered by SharePoint retention and the SharePoint recycle bin. Therefore, you can regularly export Basic plans to Excel to capture tasks, buckets, dates, and so on. You (or your power users) could automate the process with Power Automate.

Premium plans are a bit different, and you'll need to use Power Automate or perform a Dataverse data export to back up your plan data. Microsoft 365 Groups retention enables you to recover deleted plans up to 30 days after deletion. The backups that you and Microsoft make in SharePoint environments ensure that any Planner documents stored in SharePoint are backed up without requiring additional backup solutions.

The backup and recovery strategy that you choose depends on your organization, plan types, and several other considerations that make Planner backup and recovery very situational. If you aren't sure whether your existing backup and recovery strategy is sufficient, test it with some sample Groups and plans. If you find the results insufficient, consider implementing a third-party solution that specifically supports Planner. And above all, plan for the worst-case scenario (your backups are bad) — test your recovery strategy on a regular basis to avoid unwanted surprises.

The Part of Tens

Chapter **17**

Ten Tips for Faster, Smarter Planning

Everyone has their own work style, and you probably have your own ways of working in Microsoft Planner that suit your style. Even so, there are things you can do to be more effective and efficient in planning and managing plans with Planner. You can also make other people better at planning and task execution by leveraging some key features in Planner. This chapter offers ten tips to help you become a Planner rock star.

Adopt a Sprint Mentality

Sprints are an aspect of the agile project management model that break projects into time-boxed periods of one to four weeks. The sprint model is great for most types of projects because it forces you and your team to be very deliberate about defining work items, fitting them into the various sprints, and meeting regularly for checkpoints. The sprint review and retrospective at the end of each sprint inform the next sprint and help you adapt and improve from one sprint to the next. By using sprints, you can often do a better job of staying on track, identifying risks and challenges early, and adapting to changes in the project.

Make the Most of Templates

Application templates are not a new idea — they've been around in Microsoft Office apps almost since the first versions were released. With Planner, templates can save an enormous amount of time in setting up plans for specific use cases. More than that, templates help enforce a structure on your plans that tailor them to those use cases. Even if a Planner template gives you only half of what you need for a plan, use the template as a starting point to craft a custom plan. Start with the template, and add your own buckets, sprints, custom fields, and other elements to suit the project and your organization. Then save that plan as a template to use for similar future projects.

Do Some Loops

Microsoft Loop — Microsoft's real-time shared content platform — presents a nifty way to enable people to collaborate on content across many different platforms including Microsoft Teams, Microsoft Outlook, Microsoft SharePoint, and other Microsoft 365 applications. Regardless of where the Loop components live, Loop keeps everything synchronized. Best of all, people can work with the Loop components in whatever application makes the most sense at the time, be it SharePoint, Teams, Planner, Loop, or other places. If you're using Project Planner agent to execute tasks and create content, the agent creates that content as Loop components. When you build out the framework for your projects, integrate Loop in other ways to give your team more flexibility for collaboration.

Group Tasks in Multiple Ways

When you think of organizing tasks in Planner, buckets are probably the first thing that comes to mind. Buckets give you a way to organize tasks by workflow status, project phase, priority, functional area or team, work content, sprint, outcome, or deliverable. Best practice is to use buckets for the primary dimension of your plan, such as status, phase, or sprint. Then use labels, assignments, and other properties as secondary dimensions. Keep the number of buckets manageable (generally fewer than ten), and rename buckets as necessary as you work through the project. After you've built a logical method and applied it consistently across your plan, you can use the Group By option in Board view to visually organize the tasks in other ways — not just by bucket.

Use Labels

Labels are an additional property you can apply to tasks in Planner. Each unique label has a color and a description. The set of default labels that use only the color name as the label description are useful, but labels become most useful when you customize them to your plan. How you customize the labels in a plan depends on the plan type, but here are some options to consider when deciding what your custom labels will represent in the plan:

» Task type

» Risk level

» Work stream

» Stakeholder

» Skill or discipline required

» Approval or review status

» External dependency

Set up the labels before the project kicks off to ensure consistency. Build labels into your templates to create a repeatable model.

Leverage Copilot

The Project Manager agent is Microsoft's Copilot–based artificial intelligence (AI) agent for Planner. The agent can create tasks for you, execute tasks, analyze progress and risks, generate reports, and provide deep insights into your plans and projects. The agent can be extremely useful in doing research and analysis as part of a project, saving you immense amounts of time. It can incorporate company requirements, external expert sources, compliance requirements, and much more to help you build a project plan that goes a long way toward meeting internal and external requirements. Best of all, Project Manager agent can accomplish tasks in minutes that could take you days of effort. You'll need an appropriate Copilot license, but the investment is well worth it!

Use Checklists Extensively

Checklists help bring order to what could be chaos, whether it's in your daily life, a simple plan, or a complex project. Adding checklists to Planner tasks helps you define key steps that need to be performed within a task to bring the task to completion. Add as much detail as makes sense with checklists in the plan, from just adding high-level activities to detailing each step. The checklists can be particularly useful when you assign the task to someone else who may not be familiar with all the requirements or steps needed.

Checklists are a great way to remind the task owner of documentation, research, policy requirements, and other actions that they may otherwise miss. Building the checklist in a task can also help you as you create the task, giving you a cross-check to make sure you've covered all the bases.

Manage Your Personal Tasks, Too

Planner Basic can be a great tool for organizing your own personal projects. It's simple, it's easy to use, and it integrates nicely with the Microsoft To Do app and Outlook. It's also secure — other people won't see your personal plans unless you take the extra step of sharing them. When you do need help with a small project, create tasks for the other people who'll be helping and assign the tasks to them. You may also discover as you begin to use Planner more and more for your personal projects that the small added expense for a Planner Plan 1 license to add features like subtasks, Timeline view, sprints, and other Premium features is well worth the investment.

Visualize with Colors

Simplifying and bringing the most salient information into a single report page can do wonders to help people quickly find the information they need. The next step is to make that information easily consumable. Color presents a very useful means of highlighting important information, and you can use color in Planner to not only add clarity to your Planner reports but also add clarity to the various task views. For example, apply color to buckets based on each bucket's content, assigned team, or other logical criteria (colored buckets are a Premium feature). Use colored labels to highlight tasks by similar criteria that fits the project type and your organization's processes. Use color in combination with filtering and grouping to easily focus on specific tasks, assignees, and other plan elements. Equally important, document these color choices in the documentation for the project so the team understands what each color represents.

Enforce Consistency

Whenever possible, enforce consistency in your project plans. For example, always create Read Me documentation for each plan that fully explains process, color coding, outcomes, and so on for the plan, and ensure that all team members are aware of the documentation and have easy access to it. Formalize and document a color scheme that you use for all plans. Require checklists for all tasks to highlight key requirements and deliverables.

Also, require team members to input information in specific ways where possible by using Choice custom fields and clearly defining the choices you make available through those fields. This ensures higher-quality data, which leads to better reporting and greater consistency within the project plan.

Chapter **18**

Ten Outside-the-Box Use Cases for Microsoft Planner

Most applications offer flexibility in how you can use them, and Microsoft Planner is no exception. Although the general principle of Planner is to help you plan and complete projects, the types of projects and the ways in which you can use Planner to complete them are very broad.

In this chapter, I offer insight into ten creative, outside-the-box ways you can use Planner to accomplish goals at home and at work. The key to many of these use cases is to think of tasks as objects rather than activities and use Planner to manage those objects.

Managing Warranties, Services, and Expirations

In our society, we're surrounded by *stuff*. Take a minute to think of all the things that you use or rely on every day: car, computer, phone, appliances, the water filter in your refrigerator, smoke and carbon monoxide detectors, your water softener, humidifier, dehumidifier, and on and on. All these things have warranties and/or need to be serviced on a regular basis. Are you conscientious about changing the batteries in your smoke detectors? If not, here's a suggested plan that can help you get a handle on that task along with all the other stuff in your life:

>> **Buckets:** Use buckets to organize your stuff by area or type. For example, you could create a bucket for all your appliances, with each task representing one appliance. Or you could be more granular and create a bucket each for laundry, kitchen, TVs, phones, cars, and so on.

>> **Tasks and due dates:** Create a task for each appliance, device, and so on. Set the due date to the date the device needs service (such as your water softener or smoke detectors) or to the date the device's warranty expires. Back up the date a bit if you need to have time to renew it.

>> **Checklists:** Add checklists to each task to call out the actions you need to take for each item.

>> **Labels and priority:** Use colored labels and priority to highlight items that are more critical or need more lead time.

Volunteering

Whether you're managing a single event or multiple events for a group, Planner can help bring together all your people and resources to ensure success. The types of events will determine how you structure your plans, but here are some recommendations to get you started:

>> **Buckets:** Use buckets to organize your team by role (for example, create buckets for registration, facilities preparation, safety, food, outreach, coaching, and so on).

>> **Tasks:** Use tasks to assign volunteers to specific roles or activities, with checklists to define the requirements of the tasks or needed resources.

>> **Labels:** Use labels to identify skills, certifications, or other information applicable to each task. You can also use labels to indicate shifts.

The key here is to use tasks to help the volunteers understand what they need to do, when they need to do it, and what resources they'll need.

Creating and Publishing Content

Writing this book is a good example of where Planner could've been handy. I say "could've been" because I slogged through the project without it. (I'll have a different view for the next book I write.) Whether you're writing a book or article or producing some other kind of content, Planner can help you organize the project and keep track of your deadlines. Here's how you might set up the plan:

>> **Tasks:** Tasks represent content assets in this use case. For a book, each chapter and part would have its own task complete with title, description, and start and due dates. For an article or blog post, the tasks could represent thematic sections or follow a start–middle–end structure.

>> **Buckets:** One option is to use buckets to organize the content by progress stage. For example, a book could have Draft, Authoring, Development Review, Author Revision, Copyedit, and Final Review buckets. Move tasks into the appropriate buckets as you work through each element.

Microsoft 365 Copilot can really offer value in this use case. If you have a Copilot license for Planner, use Project Manager agent to perform research for you for each element of your book, article, blog post, and so on. Just make sure to check the data for accuracy before you trust its content.

Running a Home Renovation Project

Whether you're remodeling your home yourself or acting as a general contractor for other people doing the work, Planner can be a great asset for keeping the project on track and on budget. The plan will differ depending on which route you take, but here are some suggestions to get started on your renovation plan:

>> **Organize by phase.** Create buckets for the different phases of the project, such as planning, permitting, trade selection, demolition, rough-in, finishes, punch list, and inspections.

>> **Create clearly defined tasks.** Create granular tasks such as "Remove tub," "Prepare subfloor," "Tile floor," and so on. Use start and due dates to model the timeframe for the activity.

>> **Leverage checklists to ensure proper completion.** Use checklists in each task to define the requirements and steps for completing the task.

>> **Document the job.** Keep track of material selections (tile, countertops, and so on) in a folder on your local drive or in Microsoft SharePoint, and include links in the tasks to their respective documents.

Planning a Trip

My favorite kind of trip is to jump in the car and see where we end up at the end of the day, kind of like an Australian walkabout. If you're taking a long trip, traveling by air, going overseas, and so on, you need to do a bit more planning! Naturally, Planner is a great choice for planning your trip:

>> **Organize the trip by phases.** Use buckets for the different phases of the trip, such as Planning, Booking, Pre-Trip, Outbound Travel, Locations and Activities, Return Travel, and Post-Trip.

>> **Use tasks for booking and activities.** Create tasks for things like getting your passport and visa, booking airfare or cruise tickets, booking hotels, arranging activities, and other key trip items.

>> **Timing:** Use due dates for tasks like renewing your passport, applying for a visa, and booking tickets, as well as for your departure, return, and other key dates.

Optionally, use labels to highlight events or other criteria like down days, optional activities, sightseeing, in transit, and so on. Use checklists to make sure you capture all the requirements for each task, such as gathering the documentation for your passport.

Managing a Training Event

Whether you're delivering some training yourself or helping someone else organize a training event, Planner can help you develop the content, organize the training, and deliver a successful event. It works equally well for one-day events and multiday training sessions. It really shines for repeatable training events because after the framework is established, you can refine the plan based on successes and challenges from the previous session. Here are some suggestions for plan setup:

>> **Buckets:** Organize the plan by event phase including Define Objectives, Design Content, Pre-Event Logistics, Marketing, Registration, Delivery (by day), Evaluation, and Post-Event Review.

>> **Tasks:** The tasks should define discrete deliverables like Learning Objectives, Slide Deck Day 1, Book Facility, Facility Setup, Post-Event Survey, and so on. Avoid nebulous tasks like Develop Training.

>> **Dates:** Due dates should define hard milestones for completing each task.

>> **Checklists:** Use checklists to define the requirements for each task.

>> **Documents and collateral:** Have a plan to store slide decks, handouts, and other collateral in a common location, and link to the documents from the plan.

Writing a Novel

I think *Microsoft Planner For Dummies* is my 66th nonfiction book. I've sold a few short stories over the years, but I've always wanted to write a novel. Planner could be a good solution for helping me develop the plot, scenes, characters, and other book elements. A good tale follows a strong *story arc*, which is the path the author uses to tell the story. Scenes are scattered along the path like gemstones leading from some inciting event, through conflict to resolution. Here's how I would capture those gems in Planner:

>> **Plot:** Tasks in this bucket represent key events or other elements that drive the story from beginning to end. Custom fields capture characters and locations for each key event.

>> **World:** This bucket holds tasks that describe the world in which the story takes place. Tasks define the world overall and locations within the story world.

>> **Artifacts:** These are key objects in the story: a home, a starship, a magical sword, a terrible weapon.

>> **Research:** This bucket would contain tasks for Copilot to perform research, as well as research tasks I execute myself.

>> **Characters:** Tasks in this bucket represent and describe the characters in the story. Physical characteristics and motivations can be captured in the task or linked to character studies in external documents.

>> **Scenes:** These are the gems strewn along the story arc where characters, locations, and artifacts come together to paint the story. Tasks in this bucket describe the scene, identify where it falls within the story by timeline, and link out to an external document for the scene's draft.

I can't wait to get started!

Gaining a New Skill

Whether it's learning to play an instrument, learning a new trade, getting into a new hobby, or gaining a certification, Planner can help you identify key requirements, equipment, and other items that will build the necessary skills. A good approach in each case is to build a realistic timeline into the plan that considers required activities, equipment needed (including acquisition time), learning, practice, and other elements. Here are some suggestions for setting up such a plan:

>> **Do your research.** Whatever the skill or certification, research different learning paths, equipment needed, requirements to be met, and all other aspects of moving your skill level from where it is to where you want it to be. Where possible, leverage Copilot and Project Manager agent to do a lot of the research for you.

>> **Create a journey.** Armed with your research, build a plan that takes steps toward discrete, attainable goals. Build buckets that reflect that journey: Foundation, Resources, Basic Techniques, Practice Opportunities, Advanced Techniques, and Certification. Tasks in these buckets describe time-boxed activities to build from foundational skills to mastery. Focus each task on a specific learning activity and outcome.

>> **Use checklists.** Checklists can help you define the step-by-step actions needed to complete the learning task.

>> **Take stock.** Routinely evaluate your success and comfort level and use labels to flag completed tasks as mastered, needing more review, or other confidence markers to help you more easily track the mini-skills that would benefit from more experience.

>> **Keep a journal.** Create a Journal bucket where you can keep notes about key insights, lessons learned, frequent mistakes, and other findings along the way.

Tracking Progress without Meetings

I wish I had a dollar for every meeting I attended either in person or remotely over the span of my career. I'd be writing this from a lovely beach somewhere. Meetings can certainly be useful, but they can also be the bane of your existence. Wouldn't it be great to limit those meetings where you take time out from getting things done to talk about why you're not getting things done? Planner can help by providing a way for people to collaborate on a project and stay informed without excessive meetings. Here's how:

>> **Make Planner the source of truth.** All work that happens for the project happens in Planner. Avoid external documents, spreadsheets, and other resources wherever possible. When documents are needed, make them easily accessible through links to SharePoint.

>> **Build buckets with stakeholders in mind.** Think about how your project stakeholders visualize progress and success, and model the buckets using that insight. This will enable stakeholders to more easily visualize project status on their terms.

>> **Use outcomes for task titles.** Be descriptive of the task's outcome. For example, use "Initial application interface wireframes completed" instead of "Initial interface work." This helps everyone understand the significance of the tasks and its completion.

>> **Implement and communicate status signals.** Choose status indicators that make sense to the stakeholders and communicate their significance to all project participants.

>> **Develop good reports and publish them regularly.** Keep the reports reasonably short and focused on outcomes, blockers, risks, stakeholder actions, and overall progress.

>> **Escalate within Planner.** When tasks are blocked, mark them as such, add comments in the task accordingly, and notify the person or team that needs to unblock the task. Try to resolve issues outside of meetings wherever possible.

Building a Life

Most people have something about their lives that they want to improve. Maybe it's gaining physical or emotional health, improving their financial situation, being more active in the community, or improving their work/life balance.

Sometimes we never start because each one seems daunting. Pile on two or three more, and it becomes too easy to sit back and accept your current reality. Maybe you just need a plan!

Here are some suggestions for using Planner to get a handle on those things in your life that you want to change:

>> **Don't try to boil the ocean.** Create a separate plan for each goal. For example, create one for physical health, another for financial health, and another for community engagement. Don't lump them all together. You can create plans for all kinds of goals, but be realistic about how many you can work at one time. It's okay to have a plan on the back burner while you work on some others. Be flexible and allow yourself to move between goals to adapt to workload or other factors.

>> **Organize for progress.** Create buckets based on how you'll progress through the plan and not on task categories. For a physical health plan, this could include buckets for taking stock of your current situation, month-by-month activities, and evaluating your progress at a six-month milestone.

>> **Create discrete, achievable tasks.** Don't put hurdles in your path (unless you want to run hurdles in track). Create tasks that you can achieve each week to build progress and a sense of accomplishment. Create a task to lose two pounds in week one, not ten pounds.

>> **Use recurring tasks.** These are useful for things like working out and other activities that you'll repeat throughout the plan.

>> **Track and reflect within the plan.** As you complete a task, make sure to mark it as complete and use Planner's views to visualize how you're progressing overall. Keep notes or use labels to identify things you really enjoyed and things that really challenged you.

>> **Recognize and celebrate small achievements.** Each week, review and celebrate the things you completed during the week. Some weeks will see more progress than others, but recognize that every task completed drives toward your goal.

Chapter **19**

Ten Training and Support Resources

Microsoft Planner For Dummies may be your main resource for mastering Microsoft Planner, but it certainly isn't the only resource you can lever-age. In this chapter, I gather ten resources that will help you explore and master not only Planner, but related topics like Microsoft Power BI, Microsoft Power Automate, and project management in general. Whether you're new to project planning or a pro, these resources will help you fill up your box of planning tools.

Planner Help

I list this option first because it's available right from Planner! To access Planner's Help content, click the question mark in the upper-right corner of the Planner header. Doing so opens the Help pane, where you can search for content on specific features and view featured articles on Planner topics. If you don't see the topic you need in the featured articles, just click the search bar and type a description of the topic or issue. The pane will list a selection of articles that could be helpful. This is your go-to resource for getting information before branching out to other resources.

Microsoft Learn

One of my favorite sites for getting up to speed on Microsoft technologies is Microsoft Learn (https://learn.microsoft.com). Learn is Microsoft's centralized source for product documentation and learning. You'll find product documentation, application programming interface (API) references, how-to articles, and troubleshooting tips for Microsoft products and technologies. The site includes stand-alone instructional units that include text and videos; many units include knowledge checks as well. Some technologies even include sandboxes where you can try out products and skills. Just search on the keyword *Planner* on the home page of the site to find a wealth of information on Planner features, support, API references, and more.

Microsoft Support

Now and then you'll run into problems with Planner not working the way you expect it to. Maybe Planner truly isn't working as it was designed, or maybe your perception of how it should work is wrong. Whatever the case, if you have the necessary permissions in your Microsoft tenant, you can open a support case and get an engineer to troubleshoot your issue. If you manage your own personal *tenant* (the unique, secure space in the Microsoft cloud environment where your account and other data reside) or have the necessary permissions in your organization's tenant to open support cases, navigate to https://admin.microsoft.com and click the question mark in the upper-right corner of the page header. Here you can research your issue, use the Microsoft 365 Copilot–driven Support Assistant to get help, and open a support ticket with Microsoft Support.

If you don't have the necessary permissions in the tenant, reach out to your IT admin group and ask them to open a case for you.

YouTube Guided Tours and Interactive Demos

Most learning styles work well for me, whether reading, watching, or doing. But many people have a preference for learning while watching someone demonstrate. YouTube has all kinds of videos on Planner, from short how-to videos on very specific features like using Planner instead of or in conjunction with To Do, to longer videos on broader topics. For example, there are a couple of great videos on the Portfolios feature that is available starting with Planner and Project Plan 3. The presenters show you how to use the feature and offer their takes on the usefulness of Portfolios. Visit www.youtube.com and search for "Microsoft Planner" to find all these great videos.

LinkedIn Learning

LinkedIn Learning, formerly known as Lynda.com prior to Microsoft's acquisition of LinkedIn, is a great learning resource. Unlike YouTube, where pretty much anyone can post videos, LinkedIn Learning is a curated site where Microsoft contracts experts on different topics to create learning content for the site. For example, several years ago I authored a set of videos on Windows 10 shortly after it launched. LinkedIn Learning content isn't limited to Microsoft products or even to technology topics, making it a great resource to learn about a very broad range of topics. Although some free trials are available, access to content in LinkedIn Learning requires a paid subscription. If your organization doesn't provide you with a subscription, check out the LinkedIn Learning site at www.linkedin.com/learning to contact LinkedIn Sales for more information.

Tech Community

Your best resource for direct access to other Planner users, IT professionals, and the Planner development team is the Planner Tech Community. The community is a network of forums, blogs, events, and discussion boards hosted by Microsoft and focused on Planner. At the site, you'll find blogs, news and announcements, discussion forums, tips and best practices, and access to live events and demos. The community encompasses end users and administrators, offering end-user how-to content along with guidance on implementation and management. You'll find the Planner Tech Community at https://techcommunity.microsoft.com/category/planner.

Microsoft Teams Help Center

It's a safe bet that at some point you'll need some help with Microsoft Teams considering Planner's tight integration with Teams. The Teams Help Center, located at https://support.microsoft.com/en-us/teams, offers a wealth of content, including how-to articles and tutorials, troubleshooting guidance, training resources, support links, and community forums, all dedicated to Teams. Like the Planner Tech Community, the Teams Help Center offers content for end users, IT professionals, and administrators who need guidance on Teams features, implementation, and management.

Planner and Power Automate Integration

As with many Microsoft Office 365 applications, you can automate many actions in Planner using Power Automate, Microsoft's low-code/no-code automation tool. One of the best places to learn about Power Automate is the Microsoft Learn site, referenced earlier in this chapter. To go directly to the Power Automate content, navigate to https://learn.microsoft.com/en-us/power-automate. You'll find learning content at all levels, from getting started to advanced topics, release information, troubleshooting information, and lots more. When you need more information, you can turn to other resources like LinkedIn Learning, YouTube, and others. But I would start with Microsoft Learn.

Microsoft MVP Blogs

The Most Valuable Professional (MVP) title is conferred by Microsoft on experts in specific areas. MVPs are active in tech community forums, author content, and actively participate in other activities, including training, conferences, online influence, speaking, and much more. In the dim past, I was an MVP for Microsoft Outlook. You'll find that Planner MVPs are active in the Planner Tech Community and other public Planner forums and blogs, and that's a great way to learn from them. To learn more about MVPs and find Planner MVPs in particular, navigate to https://mvp.microsoft.com, click the search bar, and search for "Microsoft Planner."

Power BI

Planner offers some nice reporting features, but you can take Planner reporting to the next level with Power BI. As with Power Automate, the Microsoft Learn site is a great resource for learning what Power BI can do and how to build reports and dashboards. All the official Microsoft Power BI documentation is hosted at Microsoft Learn at https://learn.microsoft.com/en-us/power-bi. You'll find other good learning resources at LinkedIn Learning and YouTube. One of my personal favorites is the Guy in a Cube YouTube channel (www.youtube.com/guyinacube).

Index

M

Y

About the Author

Jim Boyce has written more than 65 books and hundreds of articles on a broad range of technology topics during his nearly 40-year writing career. He spent nearly 15 years at Microsoft as an individual contributor and people manager in information technology (IT) services delivery for Premier Support and Unified Support, most recently as a cloud program director and principal customer success account manager for major and strategic accounts.

Prior to Microsoft, Jim was a director and M2 leader at Xerox, where he managed two globally distributed service practices for Windows Server and collaboration tools like SharePoint. His other experiences include building inspector, full-time freelance author, internet service provider (ISP) owner, IT consultant, college instructor, and structural steel designer/drafter. In his teens, Jim helped with the family farm's cattle, haying, and cotton harvesting operations.

Jim served as the chair of his local school board for 25 years and chair of the Executive Advisory Board for the European Children Adoption Services adoption agency for 20 years. Jim is active with Special Olympics as a coach for multiple sports and a unified player for golf. His other interests include writing, cooking, camping, woodworking, construction, flying radio-controlled aircraft, and spending time with family. He has held a private pilot's license since 1994 and is currently rated for single-engine land aircraft.

Dedication

I dedicate this book to Julie, my lovely wife of 48 years, who has managed to not kick me to the curb . . . yet.

Author's Acknowledgments

Every book is a team effort that combines the skills and experiences of the author, editors, proofreaders, and production staff to bring the book to fruition. I humbly thank Elizabeth Kuball for deftly guiding the development of the book and wearing multiple hats as project manager, development editor, and copyeditor. Also many thanks to the technical editor, Guy Hart-Davis, for ensuring that the content is accurate, and to all the production staff for turning a collection of Microsoft Word documents and screenshots into a great book!

Publisher's Acknowledgments

Managing Editor: Sofia Malik

Executive Editor: Steve Hayes

Editor: Elizabeth Kuball

Senior Editorial Assistant: Hanna Sytsma

Technical Editor: Guy Hart-Davis

Production Editor: Magesh Elangovan

Cover Image: © Nuttapong punna/ stock.adobe.com

Special Help: Carmen Krikorian, Kristie Pyles